TM9-1729C

WAR DEPARTMENT TECHNICAL MANUAL

ORDNANCE MAINTENANCE

LIGHT TANK M24 CHAFFEE

AND

155-MM HOWITZER MOTOR CARRIAGE M41

TECHNICAL MANUAL

by *WAR DEPARTMENT* • *SEPTEMBER 1947*

DISCLAIMER:

This manual is sold for historic research purposes only, as an entertainment. It contains obsolete information and is not intended to be used as part of an actual operation or maintenance training program. No book can substitute for proper training by an authorized instructor.

WAR DEPARTMENT TECHNICAL MANUAL

TM 9-1729C

This manual supersedes so much of TB ORD 271, 29 March 1945, as pertains to the materiel covered in this manual

ORDNANCE MAINTENANCE

LIGHT TANK M24

AND

155-MM HOWITZER

MOTOR CARRIAGE M41

TRACKS, SUSPENSION

HULL AND TURRET

WAR DEPARTMENT • *SEPTEMBER 1947*

United States Government Printing Office

Washington : 1947

WAR DEPARTMENT

Washington 25, D. C., 9 September 1947

TM 9-1729C, Ordnance Maintenance, Light Tank M24 and 155-mm Howitzer Motor Carriage M41: Tracks, Suspension, Hull and Turret, is published for the information and guidance of all concerned.

Information in this manual is effective as of 11 June 1947.

[AG 300.7 (9 Jan 47)]

By order of the Secretary of War:

Official: DWIGHT D. EISENHOWER
EDWARD F. WITSELL   *Chief of Staff*
Major General
The Adjutant General

Distribution:

AAF (5); AGF (5); T (5); Dept (5); Arm & Sv Bd (1); Tech Sv (2); FC (1); PE (Ord O) (5); Dist 9 (3); Establishments 9 (3); Gen & Sp Sv Sch (5); Tng Ctr (2); A (ZI) (15), (Overseas) (3); CHQ (2); D (2); T/O & E 9–7 (1); 9–9 (1); 9–37 (1); 9–65 (1); 9–67 (1); 9–316 (1); 9–317 (1); & 9–325 (1).

For explanation of distribution formula, see TM 38–405.

CONTENTS

This manual supersedes so much of TB ORD 271, 29 March 1945, as pertains to the matériel covered in this manual.

CHAPTER 1

INTRODUCTION

1. Scope

a. These instructions are published for information and guidance of personnel responsible for field and base maintenance on light tank M25, and 155-mm howitzer motor carriage M41. They contain information on maintenance which is beyond the scope of the tools, equipment, or supplies normally available to using organizations. This manual does not contain information which is intended primarily for the using arm, since such information is available to ordnance maintenance personnel in 100-series TM's or FM's.

b. This manual contains a description of, and procedure for, removal, disassembly, inspection, repair, and assembly of the track suspension system, hull, hull electrical equipment, turret, howitzer mount, and spade assembly.

c. TM 9-729 contains operating and lubricating instructions for the light tank T24 (M24) and contains all maintenance operations allocated to second echelon.

d. TM 9-744 contains operating and lubricating instructions for the 155-mm howitzer motor carriage M41 and contains all maintenance operations allocated to organizational maintenance units.

e. TM 9-1729A contains a description and procedure for disassembly, cleaning, inspection, repair, and assembly of the engines, and of the components of the cooling systems and fuel systems of light tank M24, and twin 40-mm gun motor carriage M19.

f. TM 9-1729B contains the same essential information on the transmission, transfer unit, propeller shafts, controlled differential, and final drives of light tank M24, and twin 40-mm gun motor carriage M19.

g. TM 9-1731G contains service information on the hydraulic turret traversing mechanism used in these vehicles.

h. TM 9-1825A contains service information on the Delco-Remy electrical equipment used in these vehicles.

i. TM 9-1826A contains service information on the Carter carburetor used in these vehicles.

1

j. TM 9-1829A contains information on speedometers and tachometers.

k. TM 9-1731D contains information on azimuth indicators.

l. TM 9-1828A contains information on fuel pumps.

m. TM 9-1313 contains overhaul and maintenance information on the Gun, 75-mm, M6 and the mount, combination gun, M64.

2. Forms, Records, and Reports

a. GENERAL. Forms, records, and reports are designed to serve necessary and useful purposes. Responsibility for the proper execution of these forms rests upon commanding officers of all units maintaining this equipment. It is emphasized, however, that forms, records, and reports are merely aids. They are not a substitute for thorough practical work, physical inspection, and active supervision.

b. AUTHORIZED FORMS. The forms, records, and reports generally applicable to units maintaining this equipment are listed below with brief explanations of each. No forms other than approved War Department Forms will be used. Pending availability of forms listed, old forms may be used. For a current and complete listing of all forms, see current FM 21-6 (lists and index of War Department publications).

(1) *War Department Lubrication Order.* War Department Lubrication Order No. 9-729 and 9-744 prescribes lubrication maintenance for this equipment. A lubrication order is issued with each vehicle and is to be carried with it at all times. Instructions contained herein are mandatory to all users of the equipment and supersede all conflicting lubrication instructions of prior date.

(2) *WD AGO Form 9-71 (Locator and Inventory Control Card).* Except when specified otherwise by the War Department, this form will be used as a bin tag, locator card, or inventory control card by all units authorized automotive spare parts.

(3) *WD AGO Form 9-72 (Ordnance Stock Record Card).* This form is prescribed for use by ordnance maintenance and depot companies.

(4) *WD AGO Form 9-74 (Motor Vehicle Operator's Permit).* This form will be issued by commanders to all operators of vehicles who are qualified to operate the particular vehicles noted on the permit.

(5) *WD AGO Form 9-76 (Request for Work Order).* This form will be used for requesting repairs, alterations, or other type of work within or between organizations and departments.

(6) *WD AGO Form 9-77 (Job Order Register).* This form will be used to keep a chronological record of work orders.

(7) *WD AGO Form 13-1 (Automotive Disability Report of Vehicles Disabled More Than 3 Days)*. This form will be accomplished and submitted as directed in current War Department instructions.

(8) *WD AGO Form 462 (Work Sheet for Full-track and Tank-like Wheeled Vehicles)*. This form will be used for maintenance services and for all technical inspections of these vehicles.

(9) *WD AGO Form 461-5 (Limited Technical Inspection)*. This form will be used for limited technical inspections to classify vehicles as to general over-all condition.

(10) *WD AGO Form 478 (Modification Work Order and Major Unit Assembly Replacement Record and Organization Equipment File)*. This form will be kept in possession of organizational maintenance personnel and will accompany vehicles upon transfer and evacuation to higher maintenance units. It will be a record of all modifications made and exchanges of major unit assemblies. Minor repairs, parts and accessory replacements will not be recorded. In the field, where no filing facilities are available, this form will be kept in a filing jacket.

(11) *WD AGO Form 811 (Work Request and Job Order)*. This form will be used by organizational maintenance units when requesting repair by a higher repair unit.

(12) *WD AGO Form 866 (Consolidation of Parts)*. This form will be used by a maintenance company for the periodic report required by higher headquarters showing the parts and materials used and issued by the company for a given period.

(13) *WD AGO Form 867 (Status of Modification of Work Order)*. This form provides a record of the status at any time of any modification work order being performed by a maintenance shop.

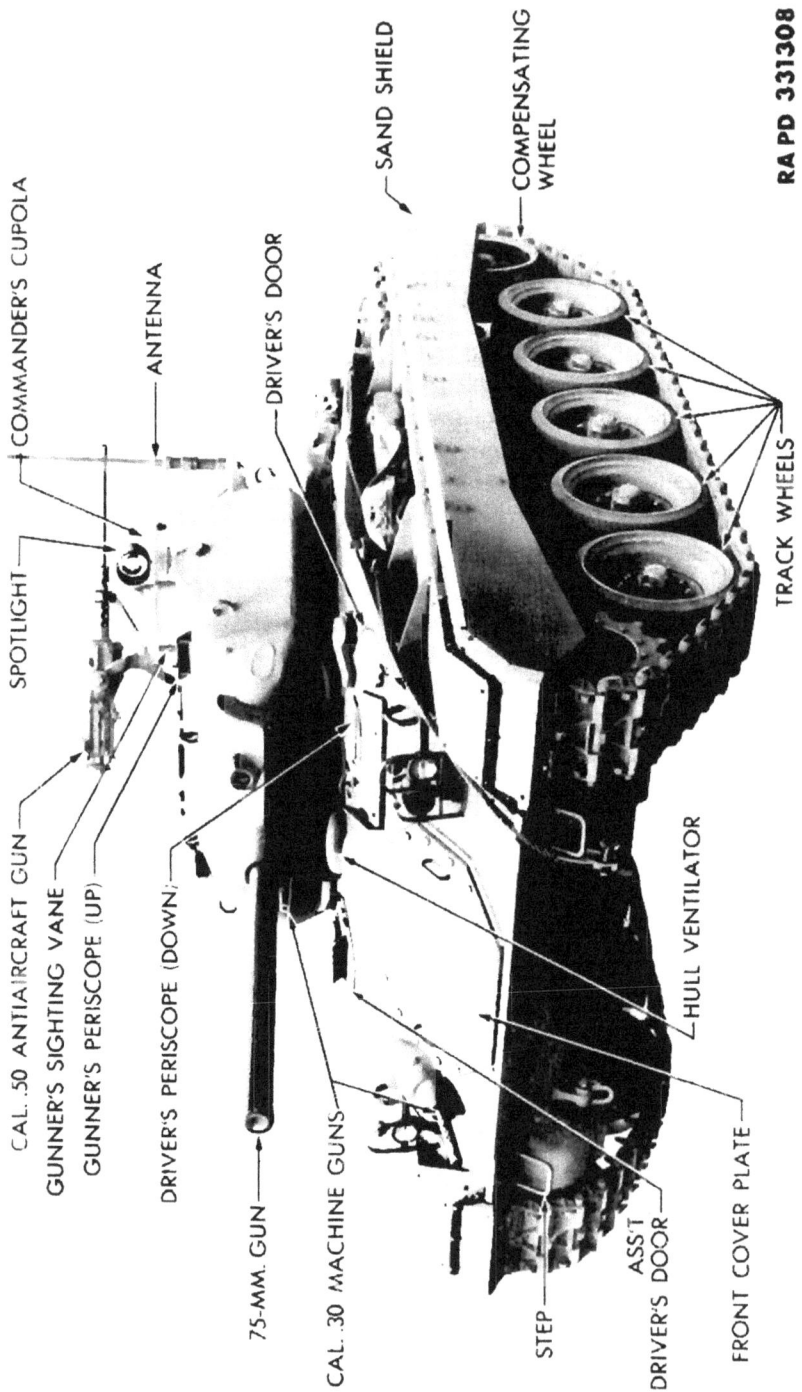

COMMANDER'S CUPOLA

ANTENNA

DRIVER'S DOOR

SAND SHIELD

COMPENSATING WHEEL

SPOTLIGHT

CAL. .50 ANTIAIRCRAFT GUN

GUNNER'S SIGHTING VANE

GUNNER'S PERISCOPE (UP)

DRIVER'S PERISCOPE (DOWN)

75-MM. GUN

CAL. .30 MACHINE GUNS

TRACK WHEELS

HULL VENTILATOR

STEP

ASS'T DRIVER'S DOOR

FRONT COVER PLATE

RA PD 331308

Figure 1. Light tank M24

4

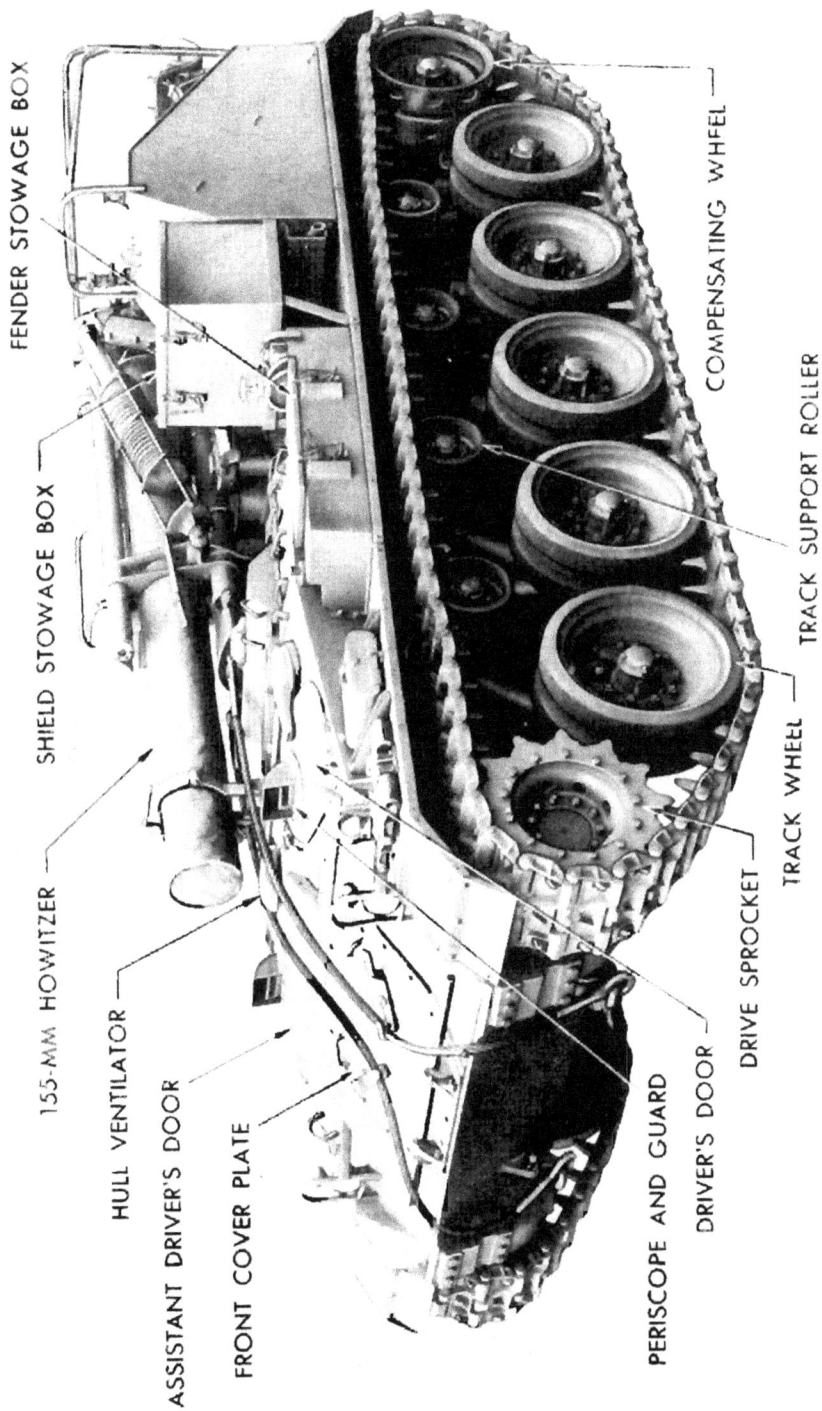

FENDER STOWAGE BOX

SHIELD STOWAGE BOX

155-MM HOWITZER

HULL VENTILATOR

ASSISTANT DRIVER'S DOOR

FRONT COVER PLATE

PERISCOPE AND GUARD

DRIVER'S DOOR

DRIVE SPROCKET

TRACK WHEEL

TRACK SUPPORT ROLLER

COMPENSATING WHEEL

RA PD 331923

Figure 2. 155-mm howitzer motor carriage M41

5

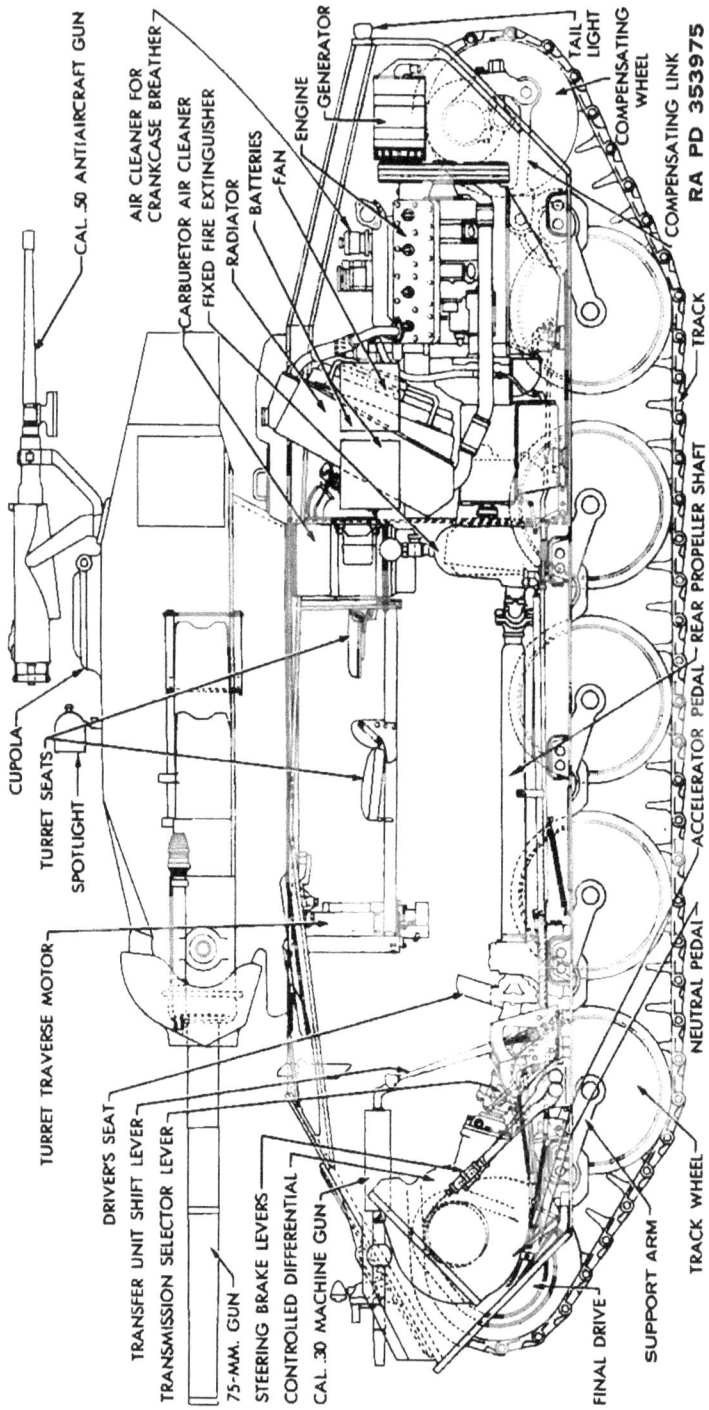

CAL. .50 ANTIAIRCRAFT GUN

AIR CLEANER FOR CRANKCASE BREATHER

CARBURETOR AIR CLEANER

FIXED FIRE EXTINGUISHER

RADIATOR

BATTERIES

FAN

ENGINE

GENERATOR

TAIL LIGHT

COMPENSATING WHEEL

COMPENSATING LINK

TRACK

REAR PROPELLER SHAFT

ACCELERATOR PEDAL

NEUTRAL PEDAL

TRACK WHEEL

SUPPORT ARM

FINAL DRIVE

CAL. .30 MACHINE GUN

CONTROLLED DIFFERENTIAL

STEERING BRAKE LEVERS

75-MM. GUN

TRANSMISSION SELECTOR LEVER

TRANSFER UNIT SHIFT LEVER

DRIVER'S SEAT

TURRET TRAVERSE MOTOR

SPOTLIGHT

TURRET SEATS

CUPOLA

RA PD 353975

Figure 3. Light tank M2₁—longitudinal section.

6

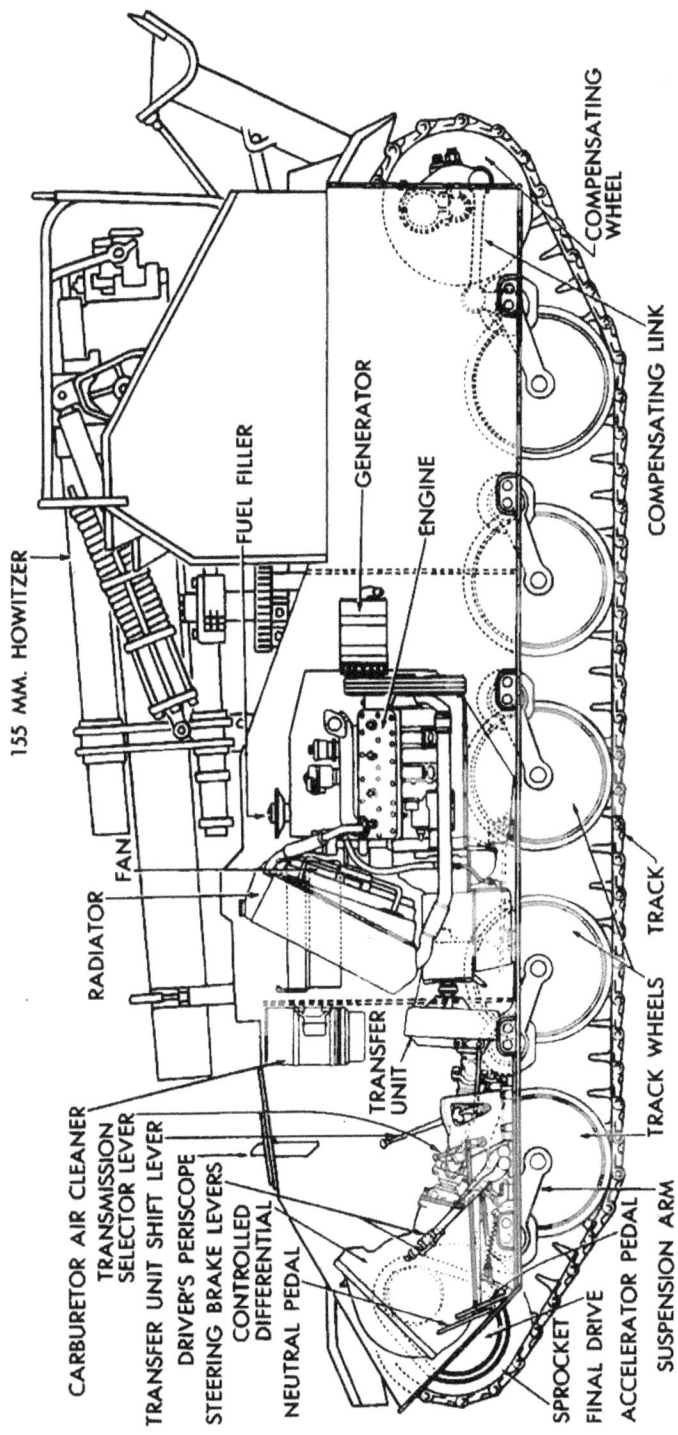

155 MM. HOWITZER

CARBURETOR AIR CLEANER
TRANSMISSION SELECTOR LEVER
TRANSFER UNIT SHIFT LEVER
DRIVER'S PERISCOPE
STEERING BRAKE LEVERS
CONTROLLED DIFFERENTIAL
NEUTRAL PEDAL

RADIATOR

FAN

FUEL FILLER

GENERATOR

ENGINE

TRANSFER UNIT

SPROCKET
FINAL DRIVE
ACCELERATOR PEDAL
SUSPENSION ARM

TRACK WHEELS

TRACK

COMPENSATING LINK

COMPENSATING WHEEL

RA PD 331926

Figure 4. 155-mm howitzer motor carriage M41—longitudinal section.

7

CHAPTER 2

DESCRIPTION

---

3. Track Suspension

a. Two individually driven steel tracks provide the means of propelling the vehicle. Each track is composed of separate steel track shoes, with integral center guides connected together with steel pins carried in rubber bushings. Two drive sprockets pull the tracks either forward or backward over the support rollers, and lay track shoes down in the path of the advancing track suspension wheels. (See figs. 5 and 6.)

b. Ten dual, rubber-tired track suspension wheels are used, five on each side, carried on individual support arms attached to torsion bars. Dual, rubber-tired track support rollers (three on M24 and four on M41) are mounted directly on the hull side walls.

c. The support arms are mounted on roller bearings in housings bolted to the hull, and are splined to a torsion bar. The torsion bars extend across the vehicle in protective tunnels just above the hull floor, and are anchored securely at their inner ends. Double-acting, airplane-type, hydraulic shock absorbers are used.

d. An adjustable compensating wheel for each track is mounted at the rear of the hull and is connected to the rear suspension arm by means of a link. This link is arranged so that any increase or decrease in track tension, due to lowering or raising of the rear suspension wheels, is offset by movement of the compensating wheel.

4. Hull

a. The hull of the vehicle is a completely welded structure, except for portions of the front, top, and floor (and rear on M41), which are removable for service operations. These removable portions consist of a plate above the controlled differential at the front of the vehicle, two drivers' doors over the drivers' seats, a large hinged door over the engine compartment, air inlet and outlet grilles for the engine compartment and radiators, and removable covers on each side, over each fuel tank and over each set of batteries. Openings in the bottom of the hull include the escape door, on M24 vehicles only, two large inspection plates (one under each engine and transmission), and the small covers just beneath the drain plugs for the engines, hydramatic transmissions, transfer unit, controlled differential, and final drives.

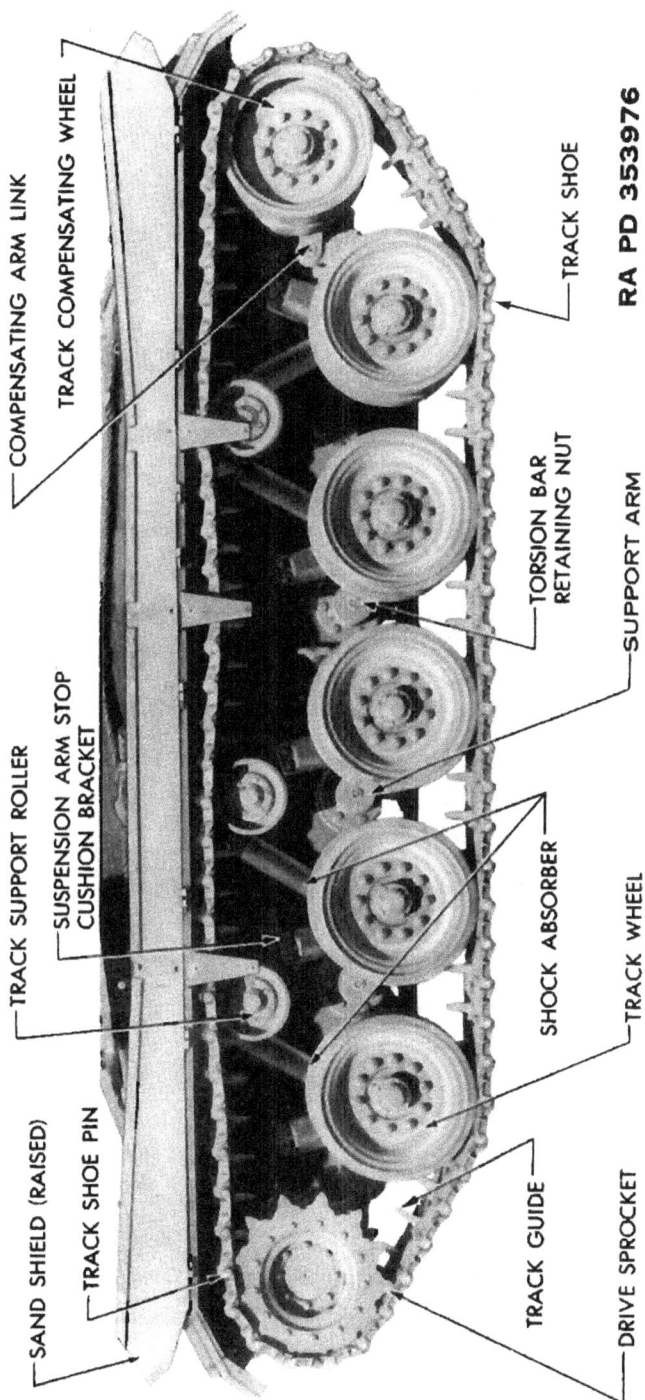

SAND SHIELD (RAISED)

TRACK SHOE PIN

TRACK SUPPORT ROLLER

SUSPENSION ARM STOP
CUSHION BRACKET

COMPENSATING ARM LINK

TRACK COMPENSATING WHEEL

TRACK SHOE

RA PD 353976

TORSION BAR
RETAINING NUT

SUPPORT ARM

SHOCK ABSORBER

TRACK WHEEL

TRACK GUIDE

DRIVE SPROCKET

Figure 5. Track suspension system (M24).

9

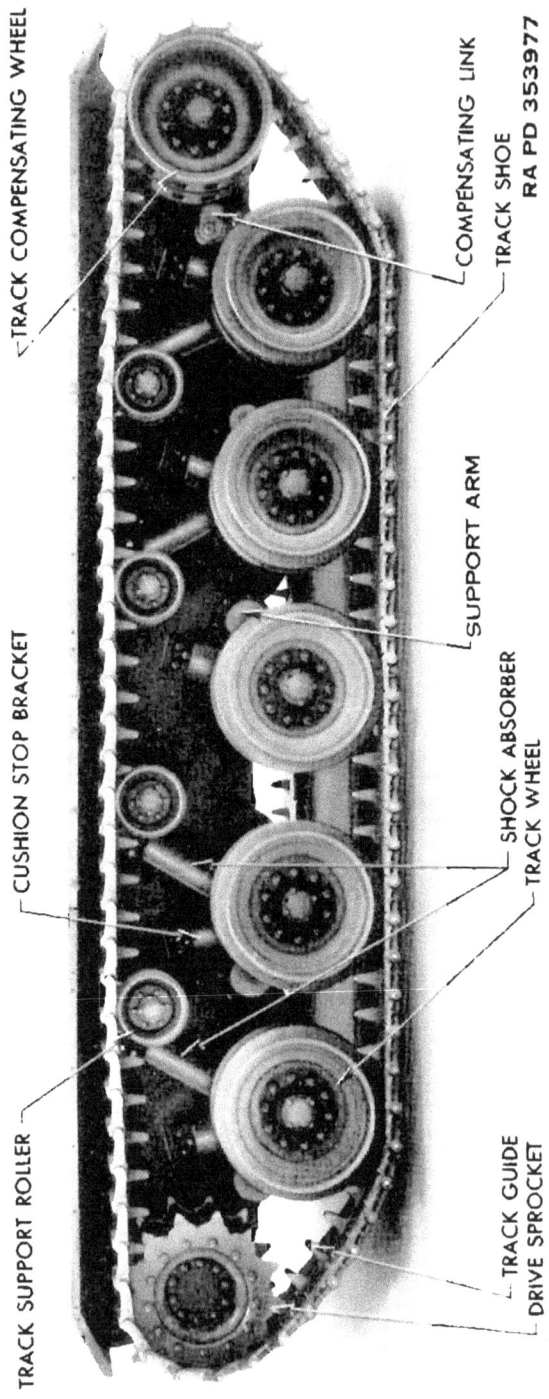

TRACK COMPENSATING WHEEL

COMPENSATING LINK

TRACK SHOE

RA PD 353977

SUPPORT ARM

CUSHION STOP BRACKET

SHOCK ABSORBER

TRACK WHEEL

TRACK SUPPORT ROLLER

TRACK GUIDE

DRIVE SPROCKET

Figure 6. Track suspension system (M41).

Figure 7. Hull (M24).

11

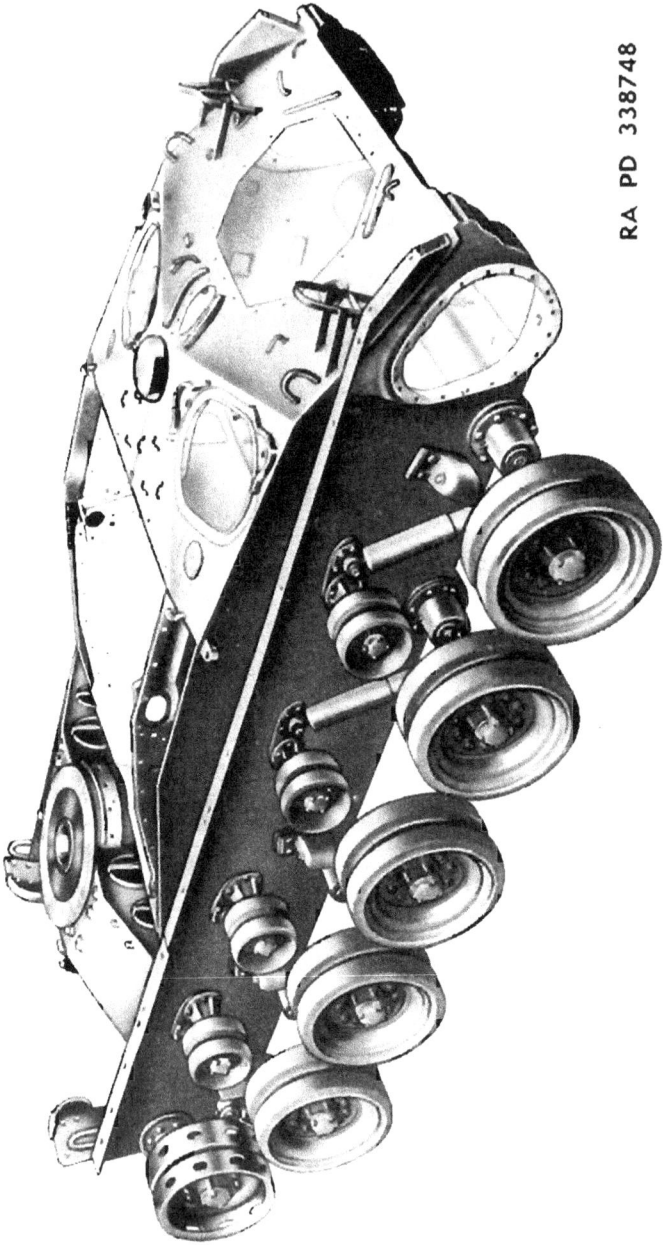

Figure 8. Hull (M41).

RA PD 338748

Figure 9. Turret assembly (M24).

b. The rear compartment of the M41 is provided with a watertight tailgate, which serves as the gunner's platform when let down in the level position over the spade during firing. Ammunition racks for 22 rounds of 155-mm ammunition are provided in the forward end of the compartment, below and at either side of the howitzer mount. Right and left gun crew shields extend from the mount to the sides of the compartment and rearward to the end of the hull.

5. Turret (M24)

a. The turret (fig. 9) is of curved armor plate, all-welded, and approximately 60 inches inside diameter. A combination gun mount, M64, carrying a 75-mm. gun, is bolted to the forward wall of the turret. The turret rotates 360° on a continuous ball bearing ring mount. This bearing is completely enclosed for protection from shell fire, lead splash, dirt, or water. The turret is traversed either by hand or by a hydraulic traversing mechanism. The turret is circular in shape, except that a bulge extends to the rear (opposite the gun). The radio is mounted in this bulge.

b. There is no turret basket, but seats for the commander, gunner, and loader are attached to the turret and rotate with it. Two hinged doors provide access to the turret, one door in the top of the commander's cupola, and the other on the right side of the turret roof, over the loaders seat.

c. Vision is provided for the commander through a periscope in the top of the cupola door, and six vision blocks at the base of the cupola. A sighting periscope for the gunner is located forward of the cupola. A port is located on the right rear side of the turret to permit discarding empty 75-mm shells with the turret closed.

Figure 10. Howitzer mount M14 on 155-mm howitzer motor carriage M41.

6. Howitzer Mount (M41 Only)

The M14 howitzer mount, in motor carriage M41, is installed in the rear compartment of the hull. It carries the 155-mm. howitzer M1 and recoil mechanism M6B1. This mount is designed to traverse 20° to the right and 17° to the left, and elevate from minus 5° to plus 45°. It is provided with an electrical elevating mechanism and control, and limit switches, in addition to the manually operated elevating mechanism. The bottom carriage is inclined forward 5°, and is welded to form an integral part of the hull of the vehicle. Except for the electrical elevating mechanism, all working parts (bearings, bushings, liners, etc.) are identical to those found in the mobile carriage M1A1, which mounts the 155-mm howitzer M1 (TM 9-1331).

CHAPTER 3

SPECIAL TOOLS

7. Purpose

a. The following list of special tools is an extract from ORD 6, SNL G–27, section 1. It contains only those special tools necessary to perform the operations described in this manual. A complete list of special tools available for all maintenance operations on light tank M24, and 155-mm. howitzer motor carriage M41 is contained in ORD 6, SNL G–27, section 1.

b. The list of special tools in paragraph 8 is for information only. It is not to be used as a basis for requisition.

8. List of Special Tools (fig. 11)

Item	Identifying number	References		Use
		Figure	Paragraph	
Sling, lifting, turret and gun assembly.	41–S–3832–54	11	37, 44	Turret replacement.
Wrench, box, single end (welded), hexagon opening, size of opening 4¼ inches, length over-all 29½ inches.	41–W–639–395	11	18	Track wheel hub replacement.
Fixture, track link assembly.	41–F–2997–392			

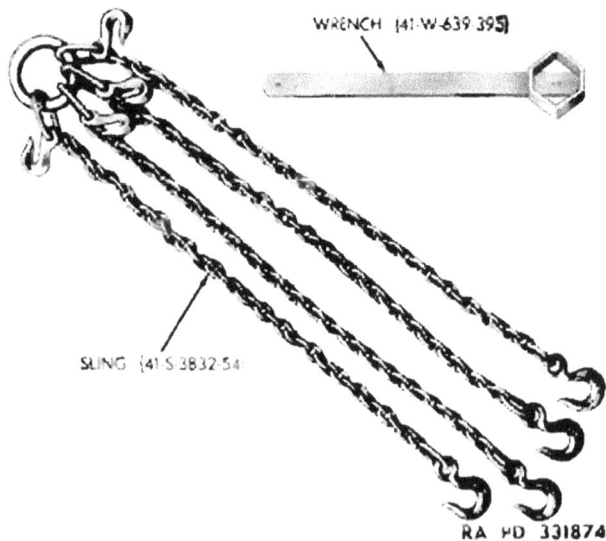

WRENCH (41-W-639-395)

SLING (41-S-3832-54)

RA PD 331874

Figure 11. Special tools.

CHAPTER 4

REMOVAL AND INSTALLATION OF MAJOR COMPONENTS

Section I. GENERAL

9. Scope

Chapter 4 contains information for the guidance of personnel in base shops, arsenals, and all other corresponding fifth echelon organizations performing major overhaul work on the light tank M24, and 155-mm. howitzer motor carriage M41. It provides an assembly line procedure for disassembly of the vehicle into major components, and assembly of the vehicle from its major components. This chapter also explains what constitutes a major component, and indicates the points of connection between major components.

Section II. DISASSEMBLY OF VEHICLE INTO MAJOR COMPONENTS

10. Disassembly

a. GENERAL. Before beginning the disassembly of the complete vehicle into major components, remove all items such as ammunition, tools, muzzle and breech covers, tarpaulins, stowage, etc. from the vehicle and tag with the hull serial number so that the same items can be reassembled on the same vehicle.

b. PROCEDURE. (1) *Remove howitzer and mount* (M41 only). Refer to paragraph 45.

(2) *Remove spade and gunners' platform assemblies* (M41 only). Refer to TM 9-744.

(3) *Remove tracks.* Refer to TM 9-729 or TM 9-744.

(4) *Remove turret assembly* (M24 only). Refer to paragraph 37.

(5) *Remove turret control box* (M24 only). Remove conduit guard from rear of turret control box and disconnect two conduits from collector ring. Remove screws holding control box to propeller shaft panel support bracket and lift out control box and collector ring.

(6) *Remove howitzer elevating control relay box and electrical conduit from gun crew compartment* (M41 only). Remove screws and clips holding electrical conduit to left and right gun crew shields (fig. 13). Remove screws holding left gun crew light to gun crew shield. Disconnect spotlight cable from outlet on spotlight bracket. Remove

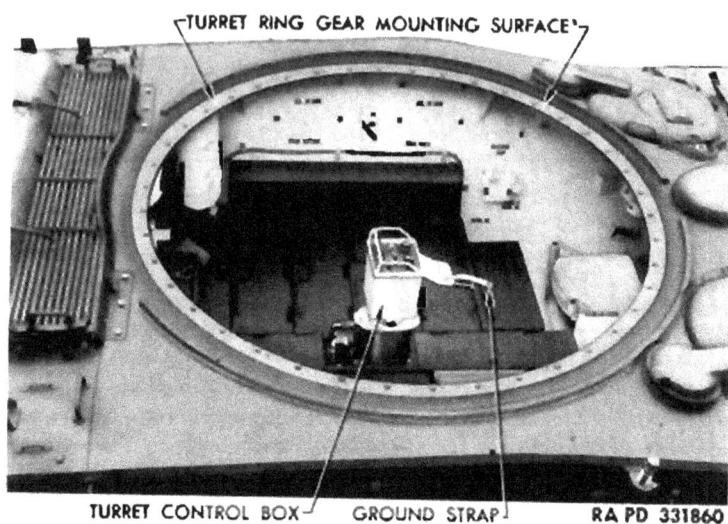

TURRET RING GEAR MOUNTING SURFACE

TURRET CONTROL BOX   GROUND STRAP   RA PD 331860

Figure 12. M24 with turret removed.

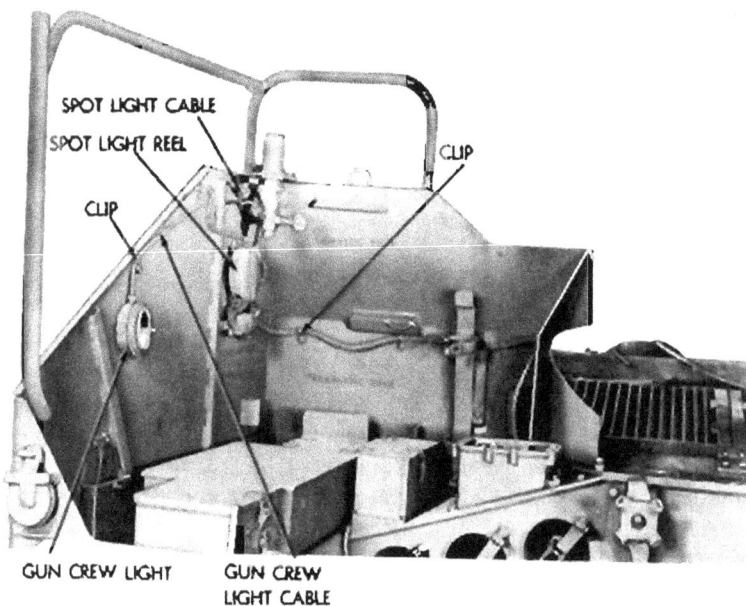

SPOT LIGHT CABLE
SPOT LIGHT REEL
CLIP
CLIP

GUN CREW LIGHT   GUN CREW
LIGHT CABLE

RA PD 338682

Figure 13. Gun crew light cables on left gun crew shield (M41 only).

18

cap screws holding spotlight reel and spotlight outlet to left gun crew shield. Disconnect the battery cable in elevating control relay box (fig. 14). Remove four cap screws and lift elevating control relay box and left conduit from gun crew compartment. Remove two screws holding right gun crew light to stowage compartment. Remove five cap screws and guard plate from inside front of right gun crew shield. Disconnect and remove three conduits from terminal box in right stowage compartment (fig. 15). Disconnect and remove conduits from right and left taillights and remove the taillights from shields.

(7) *Remove stowage boxes* (M41 only). Remove all stowage, grenade, and part boxes from gun crew compartment. Remove camouflage net and tarpaulin storage boxes from top of vehicle. Remove cap screws holding exhaust pipe guards to left and right gun crew shield.

(8) *Disconnect siren* (M41 only). Remove battery compartment cover and fuel tank compartment cover from right side of vehicle. Disconnect siren cable in rear of fuel tank compartment.

(9) *Remove left and right gun crew shields* (M41 only). Remove bolts, nuts, and lock washers holding right and left spade latches to

GUN CONTROL RELAY BOX

BATTERY CABLE   RA PD 338684

Figure 14. Battery cable to gun control relay box (M41 only).

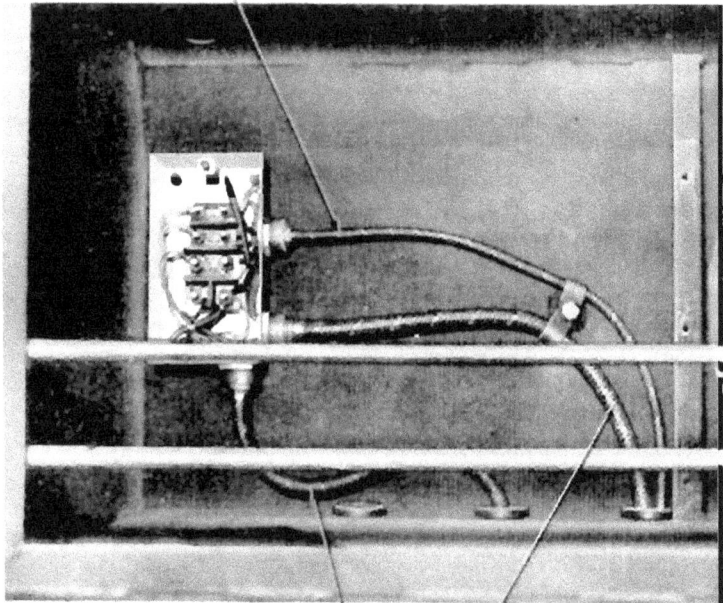

BATTERY CABLE   MAIN CONDUIT   RA PD 338683

Figure 15. Terminal box in right gun crew shield (M41 only).

gun crew shields. Remove cap screws holding left and right gun crew shields to hull and remove the shields.

(10) *Remove periscope head boxes and oddment boxes.* Remove screws and washers holding periscope head boxes and oddment boxes to propeller shaft support brackets, and lift out boxes.

(11) *Remove bulkhead doors.* Release latches from top of bulkhead doors. Tilt doors forward and lift out of channel at bottom of fire wall.

(12) *Remove bulkhead extension cover.* Remove screws from bulkhead extension cover and lift out cover.

(13) *Remove 75-mm ammunition covers and ammudamp cans* (M24 only). Remove hull subfloor from mounting brackets. Remove ammunition covers. Slide ammudamp cans out of channels to ammunition racks.

(14) *Remove driver's doors.* Refer to TM 9-729, or TM 9-744.

(15) *Remove hull top covers.* Remove the hull top covers in the following order: Differential cover plate, engine compartment door, air outlet grille, air inlet grille, radiator cover, fuel compartment covers, and battery compartment covers.

(16) *Remove batteries.* Make sure battery master switch is in the "OFF" position. Remove connecting cables from positive and negative posts of batteries. On M24 vehicle, remove center connecting strap. Remove battery hold-downs and lift out batteries.

(17) *Remove ammunition racks* (M24 only). Remove the screws and washers which hold these racks to hull side wall and propeller shaft support bracket, and then remove the racks from the hull floor.

(18) *Remove instrument panel.* Remove bolts holding instrument panel to mounting brackets. Slide panel away from front deck and remove all conduits and cables from rear of panel.

(19) *Remove differential oil cooler.* See TM 9-729, or TM 9-744.

(20) *Remove radiators.* Refer to TM 9-729, or TM 9-744.

(21) *Remove air cleaners and pipes.* See TM 9-729, or TM 9-744.

(22) *Remove fuel lines from both fuel tanks to both engines.* Disconnect cables from front of fuel shut-off valves. Remove hose clamps connecting fuel shut-off valves to fuel tank. Slide hose off connections. Disconnect both gasoline lines at carburetors. Remove screws holding roof cross bar to inner side wall and lift out cross bar with fuel lines attached. Remove fuel shut-off valves.

(23) *Remove track sprockets.* Refer to TM 9-729, or TM 9-744.

(24) *Remove final drives.* Refer to TM 9-729, or TM 9-744.

(25) *Remove driver's and auxiliary driver's seats.* Raise seat to UP position. Remove screws and washers holding seat assembly to hull floor and remove seat.

(26) *Remove braking and steering controls.* Refer to TM 9-729 or TM 9-744.

(27) *Remove differential assembly and main propeller shaft.* Refer to TM 9-729 or TM 9-744 and follow procedure given there, except that it is not necessary to disconnect propeller shafts from differential, as these units can be removed as an assembly through opening in front of hull, if desired.

(28) *Remove fixed fire extinguisher cylinder.* Refer to TM 9-729 or TM 9-744.

(29) *Remove engine assemblies.* Refer to TM 9-729, or TM 9-744.

(30) *Remove lights and siren.* Remove headlights, taillights, siren, and siren guard from hull.

(31) *Remove ventilator assembly and emergency ignition switch.* Disconnect conduits leading to ventilator assembly. Remove screws holding ventilator to hull front deck and remove ventilator. Disconnect conduits to emergency ignition switch. Remove screws holding emergency ignition switch to hull front deck and remove switch.

(32) *Remove fuel tanks.* Refer to TM 9-729 or TM 9-744.

(33) *Remove transfer unit.* Remove the three transfer unit mounting screws. Attach a rope sling around transfer unit and lift transfer unit out of vehicle through back of engine compartment.

(34) *Remove driver's door mat.* Refer to TM 9–729, or TM 9–744.

(35) *Remove hull floor plates.* Refer to TM 9–729, or TM 9–744.

(36) *Remove track suspension parts.* After blocking up hull, remove track support rollers, track wheels, torsion bars, torsion bar anchors, support arms, shock absorbers, cushion stop brackets, compensating wheels and compensating arm and lever assemblies, according to procedure in TM 9–729, or TM 9–744.

Note. Electrical conduits, control linkage parts, fire extinguisher lines, mounting, or stowage brackets should be left in the hull, unless inspection indicates a need for replacement of any of these parts.

Section III. ASSEMBLY OF VEHICLE FROM MAJOR COMPONENTS

11. Assembly

a. GENERAL. Before proceeding with the assembly of the vehicle, make any needed repairs to electrical conduits, linkage parts, or mounting parts.

b. ASSEMBLY PROCEDURE. (1) *Repair and paint hull.* Clean and paint the complete hull, inside and outside. Inspect all tapping blocks having threaded screw holes for stripped threads. Raise hull approximately 24 inches off floor and block securely.

(2) *Install track suspension parts.* See TM 9–729 (or TM 9–744.

(3) *Lubricate suspension system complete.* Lubricate complete suspension system with seasonal grade lubricant according to lubrication orders.

(4) *Install hull floor plates.* Refer to TM 9–729, or TM 9–744.

(5) *Install driver's door pads.* Refer to TM 9–729, or TM 9–744.

(6) *Install transfer unit.* Attach a rope sling around transfer unit and lower through engine compartment into position over transfer unit mounting support. Install the two front and one rear mounting screws holding transfer unit to support. Tighten screws to 45–50 foot-pounds.

(7) *Install fuel shut-off pipes and cables.* Install fuel shut-off pipes and cables under hull roof and connect front end of cable to fuel pump switch. Fasten cables to hull roof with attaching clips.

(8) *Install emergency ignition switch.* Position emergency ignition switch over tapping blocks on hull roof and install the attaching screws and washers. Connect conduits to emergency ignition switch and tighten knurled nuts on conduits securely.

(9) *Install ventilator assembly.* Position ventilator assembly over tapping blocks on hull front deck and install attaching screws and washers. Connect electrical conduits to ventilator and tighten knurled nuts securely.

(10) *Install fuel tanks and insulating pads.* Refer to TM 9–729, or TM 9–744.

(11) *Install lights and siren.* Install siren guard to hull, then install siren, taillights, and headlights.

(12) *Install engine assemblies.* Refer to TM 9–729, or TM 9–744.

(13) *Install fixed fire extinguisher cylinder.* Refer to TM 9–729, or TM 9–744.

(14) *Install braking and steering controls.* Refer to TM 9–729, or TM 9–744.

(15) *Install controlled differential and main propeller shafts.* Refer to TM 9–729, or TM 9–744.

(16) *Install drivers' seats.* Position the driver's and assistant driver's seat over tapping block on hull floor, and install the four attaching screws and washers, holding seat to floor.

(17) *Install fuel line system to right and left engine from right and left fuel tank.* Place roof cross bar in position over engine compartment and install the attaching screws and washers. Place fuel shutoff valve in position on hull side wall and install the attaching screws and washers. Install hose connection from fuel shut-off valve to elbows on fuel tanks. Tighten hose clamp securely. Connect hose from fuel line to carburetor. Tighten hose clamp securely. Connect cables from fuel shut-off valves to shut-off valve control. Install locking wires.

(18) *Install air cleaner system.* Refer to TM 9–729, or TM 9–744.

(19) *Install right and left radiators.* See TM 9–729, or TM 9–744.

(20) *Install differential oil cooler.* See TM 9–729, or TM 9–744.

(21) *Install right and left battery assemblies.* Refer to TM 9–729, or TM 9–744.

(22) *Adjust steering brakes.* Refer to TM 9–729, or TM 9–744.

(23) *Install instrument panel.* Connect all conduits and cables to front side of instrument panel and tighten knurled nuts on cable securely. Position instrument panel against mounting brackets on hull front deck, and install the four attaching screws and washers.

(24) *Install ammunition racks* (M24 only). Position ammunition racks on hull floor, and install the attaching screws and washers holding ammunition racks to hull side wall and propeller shaft support bracket.

(25) *Install hull top covers.* Install hull top covers in the following order: Fuel compartment covers, battery compartment covers, radiator cover, engine compartment cover and grilles, and differential cover.

(26) *Install final drives.* Refer to TM 9–729, or TM 9–744.

(27) *Install track sprockets.* Refer to TM 9–729 or TM 9–744.

(28) *Install right and left driver's doors to hull.* Refer to TM 9–729 or TM 9–744.

(29) *Install Ammudamp cans and 75-mm. ammunition covers* (M24 only). Place the liquid containers in the slots provided in the ammunition racks. Install the 75-mm. ammunition covers over the racks, and install the screws holding the ammunition compartment doors to the ammunition racks. Install subflooring.

(30) *Install bulkhead extension.* Place the bulkhead extension in position over the top of the transfer unit, and install the attaching screws holding extension to supports.

(31) *Install bulkhead doors.* Hook the lower part of the bulkhead door in the channel at the base of the door opening, and raise door so that the bulkhead door latches catch on the knobs at the top of the bulkhead door.

(32) *Install spare periscope head and ration boxes in propeller shaft bracket.* Place periscope and spare head, and ration boxes in position on propeller shaft support brackets and fasten securely with four cap screws.

(33) *Install left and right gun crew shields* (M41 only). Position left and right gun crew shields on hull, and secure with cap screws and lock washers. Install right and left spade latches to gun crew shields.

(34) *Connect siren.* Connect siren cable in rear of fuel tank compartment. Install battery compartment cover and fuel tank compartment cover on right side of vehicle.

(35) *Install stowage boxes* (M41 only). Install exhaust pipe guards on left and right gun crew shields. Position camouflage net and tarpaulin stowage boxes on top of vehicle and secure with nuts and bolts. Install all stowage, grenade, and part boxes in gun crew compartment.

(36) *Install howitzer elevating control relay box assembly and electrical conduit in gun crew compartment* (M41 only). Install the right and left taillight on the gun crew shields and connect the taillight conduit. Install and connect the three conduits to the terminal box in right stowage compartment. Position guard plate inside front of right gun crew shield and secure with five cap screws. Secure right gun crew light to stowage compartment with two screws. Position elevating control relay box on left side of bottom carriage and secure with four cap screws. Connect the battery cable in elevating control relay box. Secure spotlight reel and spotlight outlet to left gun crew shield with cap screws and lock washers. Connect spotlight cable to outlet on spotlight bracket. Secure left gun crew light to gun crew shield with two screws. Install all screws and clips holding electrical conduit to left and right gun crew shields.

(37) *Install collector ring and turret control box* (M24 only). Position collector ring and turret control box assembly over studs on mounting brackets. Position oddment box and periscope box on

24

respective mounting studs and install four nuts and lock washers. Tighten the nuts securely. Reconnect the two conduits to the turret collector ring. Reinstall ground strap between control box arm and conduit channel. Install conduit guard to hull floor.

(38) *Install turret assembly* (M24 only). Refer to paragraph 44.

(39) *Install gunners' platform and spade assemblies* (M41 only). Refer to TM 9-729.

(40) *Install howitzer and mount* (M41 only). Refer to chapter 8, section II.

(41) *Install tracks.* Refer to TM 9-729, or TM 9-744.

(42) *Inspect vehicle.* At this point, inspect the complete vehicle for correct operation of all instruments, light, turret traversing, etc. The vehicle is now ready for road testing.

12. Inspection

Perform a technical inspection as prescribed in WD AGO Form No. 462 and outlined in TM 9-729, or TM 9-744, monthly (or after every 100 hours) as an organizational preventive maintenance service.

CHAPTER 5

TRACK AND SUSPENSION

Section I. DESCRIPTION AND DATA

13. Description and Operation

a. TRACKS. Two individually driven, light weight steel tracks, 16 inches wide are used to propel the vehicle. Each track is composed of separate track shoes (fig. 17), made of cast steel with integral center guides, and connected together with straight steel hinge pins carried in rubber bushings. On M24 vehicles, the track has 75 shoes; on M41 vehicles, 80 shoes are used.

b. DRIVE SPROCKETS. Two drive sprockets, one on each side at the front, pull the tracks forward over the supporting rollers, and lay them down in the path of the advancing suspension wheels (figs. 5 and 6).

c. TRACK WHEELS. Ten dual, rubber-tired, track suspension wheels are used, five on each side, mounted on double, opposed, tapered roller bearings, and carried on individual support arms attached to torsion bars. On M24 vehicles, the front four wheels on each side trail the support arms, and the fifth wheels lead the arms (fig. 5), while on M41 vehicles, the first two wheels on each side trail, and the rear three on each side lead (fig. 6).

d. SUSPENSION. The support arms are each mounted in two large roller bearings (fig. 23), carried in a housing bolted to the hull sides just above the floor level, and splined to a torsion bar. The torsion bars extend across the vehicle in protective tunnels on the hull floor, and are anchored securely at their inner ends. The bars for opposite right and left wheels are carried in the same tunnel, with one bar directly behind the other, in order to permit carrying all the wheels at the same level.

e. SHOCK ABSORBERS. Double-acting, airplane-type, hydraulic shock absorbers are used at the two front and two rear support arms on each side, attached directly to the hull side walls. In addition, cushion stop brackets, in which volute springs are carried, are provided for each support arm.

j. COMPENSATING WHEEL. An adjustable compensating wheel for each track is mounted at the rear of the hull, and connected to the rearmost support arm by a link (fig. 16), so arranged that any increase in track tension due to lowering of the rear suspension wheels is offset by forward movement of the compensating wheel, and any decrease in track tension due to raising of the rear suspension wheels is offset by rearward movement of the compensating wheel.

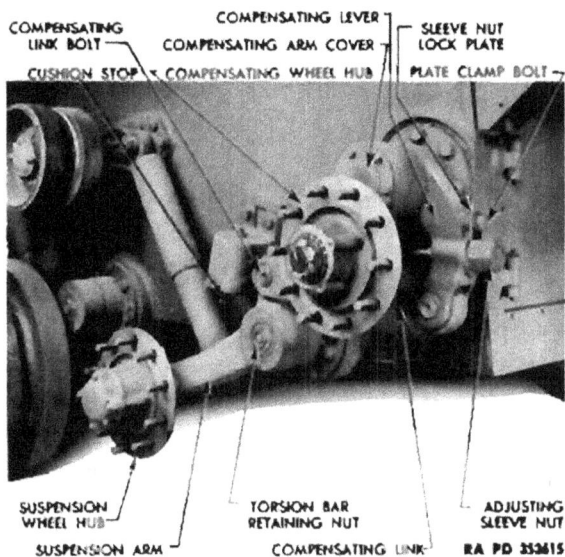

Figure 16. Compensating wheel linkage (M41).

g. SUPPORT ROLLERS. Dual, rubber-tired, track support rollers are provided to support the upper part of the track as it moves forward. The rollers are carried on double, opposed, tapered roller bearings, and mounted on support brackets bolted directly to the hull side walls. Six rollers, three on each side, are provided on M24 vehicles, and eight rollers, four on each side, are used on M41 vehicles (figs. 5 and 6).

14. Data

a. TRACK SHOES.

Number per track, M24_____ 75
Number per track, M41_____ 80
Width_____ 16 inches
Type_____ Steel, with center guide
                                         and rubber bushings.

b. TRACK.

Pitch_____ 5½ inches
Ground contact:
    Zero penetration_____122 inches
    1-inch penetration_____126 inches

c. SUPPORT ARMS.

Number (each side) _____ 5.
Type_____ Solid steel.
Type of springing_____ Torsion bar.
Number of bearings_____ 2 (straight-roller).

d. TRACK WHEELS.

Number (each side) _____ 5.
Type_____ Dual, demountable, rub-
                                                 ber-tired.
Number of bearings (each wheel)_____ 2 (tapered, roller).

e. TORSION BARS.

Number (each side) _____ 5.
Type_____ Solid steel.

f. TRACK SUPPORT ROLERS.

Number (each side), M24_____ 3.
Number (each side), M41_____ 4.
Type_____ Dual, demountable, rub-
                                                 ber-tired.
Number of bearings_____ 2 (tapered, roller).

g. SHOCK ABSORBERS.

Number (each side) _____ 4.
Type_____ Hydraulic, airplane-
                                                 type.

h. SUPPORT ARM CUSHION STOPS.

Number (each side) _____ 5.
Type_____ Volute spring.

Section II. OVERHAUL OF SUBASSEMBLIES

15. Track

a. DISASSEMBLY. Remove and disassemble track (TM 9–729 or
TM 9–744).

b. CLEANING. Wash all parts thoroughly in dry cleaning solvent
and dry with compressed air. **Caution:** *Do not allow any track parts
containing rubber to soak in solvent.*

c. INSPECTION.

Note. The specifications given in this paragraph are manufacturer's limits
on new parts.

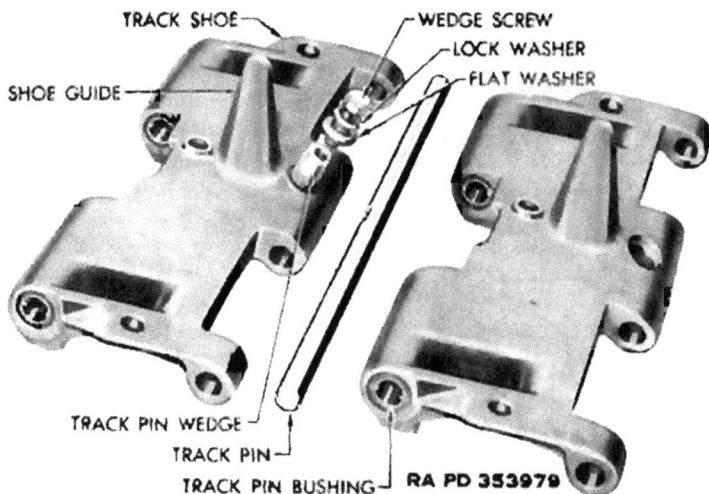

Figure 17. Track shoe disassembled.

(1) *Track pin.* Check track pin for wear and bends. Track pin must be straight for full length, within 0.004 inch total indicator reading. Diameter of pin should be 0.827 to 0.828 inch.

(2) *Track shoe.* Check track shoe for breaks, cracks, and wear. If guide lug is worn so thin that there is a possibility of it breaking off in service, discard the track shoe. Original grouser height as measured from the bushed mounting hole should be seven-eighths inch. The minimum height for overseas use is three-fourths inch, but for domestic use seven-sixteenths inch is considered minimum before replacement. Diameter of track pin opening should be 0.829 to 0.835 inch. Diameter of bushing end of shoe should be 1.365 to 1.370 inches. Thickness of a new shoe is 2¼ inches. Replace if worn beyond 1⅞ inches. Inside diameter of track pin bushing should be 0.830 to 0.832 inch.

d. Assembly. Assemble, install, and adjust track (TM 9-729 or TM 9-744).

16. Compensating Wheel, Bearings, and Seals (fig. 18)

a. Removal. Remove and disassemble compensating wheel, bearings, and seals (TM 9-729 or TM 9-744).

b. Cleaning. Wash all parts thoroughly in dry cleaning solvent and dry with compressed air.

c. Inspection.

Note. The specifications given in this paragraph are manufacturer's limits on new parts.

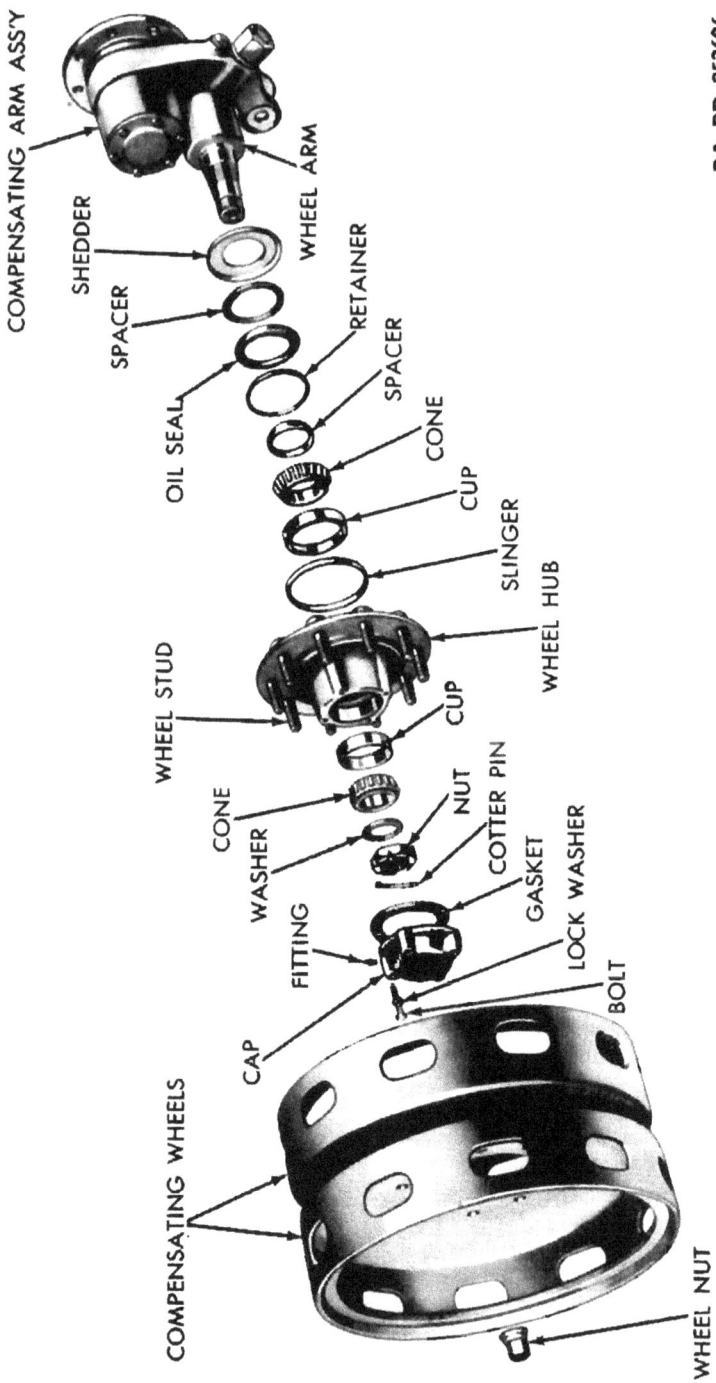

Figure 18. Compensating wheel, hub, and bearings, disassembled.

RA PD 353606

30

(1) *Compensating wheel disk.* Inspect compensating wheel disk for bends or cracks. Inspect inner sides of wheel disks for wear caused by track guides. Inspect mounting holes for wear. Diameter of holes should be 0.776 to 0.786 inch.

(2) *Hub assembly.* Inspect compensating wheel hub assembly for cracks or breaks. Inspect oil seal counterbore for grooves caused by oil seal. Inspect for stripped studs. Inspect inner and outer bearing cones and cups for wear that may cause bearing failure. Outside diameter of outer cup should be 3.6718 to 3.6728 inches. Diameter of outer cup seat in hub should be 3.6695 to 3.6710 inches. Fit of cup in seat should be 0.0008 to 0.0033 inch (tight). Outside diameter cup should be 4.4375 to 4.4385 inches. Diameter of inner cup seat in hub should be 4.4345 to 4.4365 inches. Fit of cup in seat should be 0.001 to 0.004 inch (tight).

d. ASSEMBLY. Assemble and install compensating wheel, bearings, and seals (TM 9–729 or TM 9–744).

17. Compensating Wheel Support Lever Arm, Spindle, and Link

a. DISASSEMBLY. Remove and disassemble compensating wheel support lever arm, spindle, and link (TM 9–729 or TM 9–744).

b. CLEANING. Wash all parts thoroughly in dry cleaning solvent and dry with compressed air.

c. INSPECTION.

Note. The specifications given in this paragraph are manufacturer's limits on new parts.

(1) *Compensating wheel support lever arm spindle.* Inspect compensating wheel support lever arm spindle (fig. 19) for cracks or breaks. Inspect bearing seats on spindle for scores. Diameter of spindle for roller bearing should be 2.9982 to 2.9992 inches. Diameter of spindle for ball bearing should be 2.3604 to 2.3614 inches. Inspect snap ring groove on spindle for nicks or burs. Remove small nicks and burs with a fine file. Inspect roller and ball bearings for damage or roughness that may cause failure. Outside diameter of roller bearing should be 4.4992 to 4.5000 inches. Inside diameter of roller bearing should be 2.9994 to 3.000 inches. Outside diameter of ball bearing should be 4.3301 to 4.3307 inches. Inside diameter of ball bearing should be 2.3616 to 2.3622 inches.

(2) *Compensating wheel support arm assembly.* Inspect compensating arm assembly (fig. 19) for cracks or breaks. Inside diameter of roller bearing should be 4.5005 to 4.5020 inches. Inside diameter of ball bearing should be 4.3310 to 4.3325 inches.

(3) *Compensating wheel support arm cover.* Inspect compensating wheel support arm cover for cracks or breaks. Lay cover on surface plate to check it for warp. Cover should be flat within 0.003 inch measured with a feeler gage.

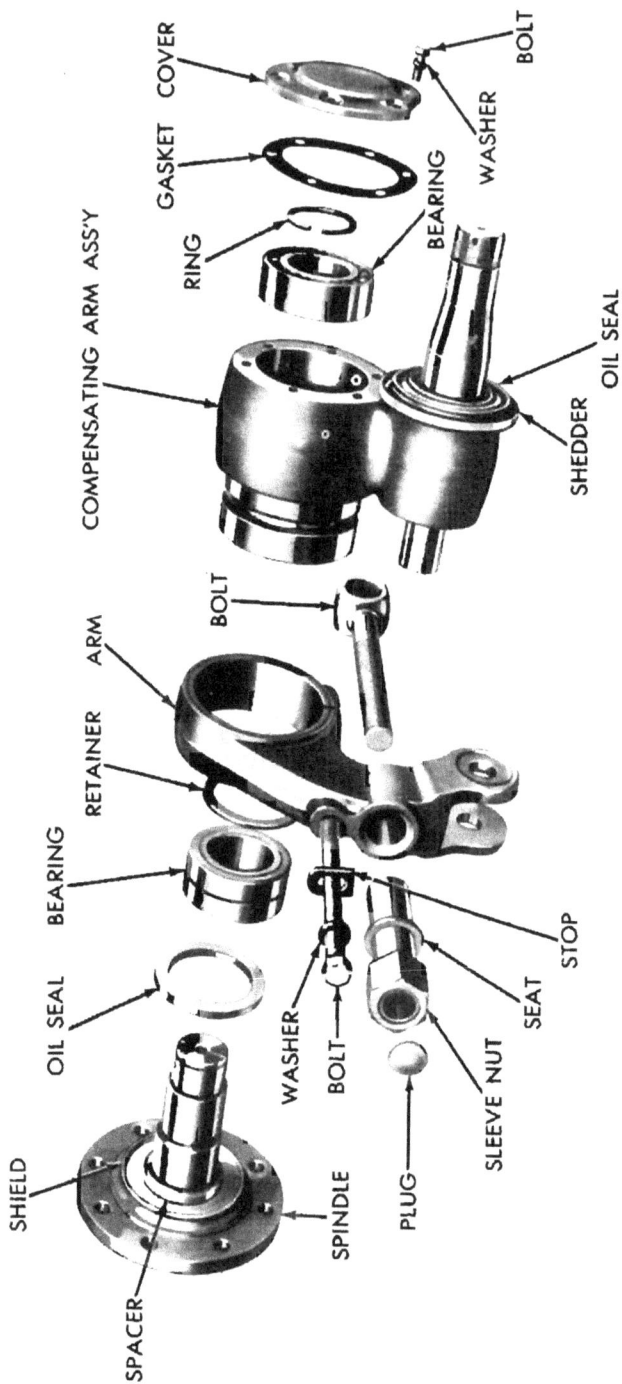

Figure 19. Compensating wheel arm and levers, disassembled.

RA PD 331904

(4) *Compensating link.* The bearings in the compensating link can be inspected by noting the position of the bolt hole in the bearing. If the bolt hole is more than one-eighth inch off center from the hole in the link, replace the bearing by pressing out the old bearing and installing a new one. Inside diameter of bearing in link should be 3.183 to 3.185 inches. Outside diameter of bearing should be 3.190 to 3.194 inches. Diameter of link bolt should be 1.1225 to 1.1235 inches. Inside diameter of bearing should be 1.1245 to 1.1255 inches.

d. ASSEMBLY. Assemble and install compensating arm support, spindle, and link (TM 9-729, or TM 9-744).

18. Track Wheels, Hubs, and Support Arm and Housing Assemblies (fig. 22)

a. REMOVAL AND DISASSEMBLY. (1) *Track wheels and hubs.* Remove and disassemble track wheels and hubs (TM 9-729 or TM 9-744).

(2) *Support arm position test (in vehicle).* Torsion bars and component parts may be checked to determine serviceability by the following method:

(*a*) With track wheels, hubs and shock absorbers removed, check backlash or free movement of support arms. Backlash, measured at the arc of the wheel spindle, should not exceed three-eighth inch.

(*b*) Raise support arm by hand to take up backlash and check distance from center of spindle to center line of track support rollers. If distance is less than that shown in figure 20, the torsion bar may be distorted or support arm may be bent. If installation of a serviceable torsion bar does not correct the condition, the support arm must be replaced.

(3) *Support arm and housing.* Remove seven screws that hold support arm assembly to hull side and lift or hoist arm and housing assembly out of hull opening. Place assembly in vise and remove large nut holding housing to arm assembly, using suspension arm bearing nut wrench (41-W-639-395), (fig. 21). Pry inner oil seal out of housing. Remove inner oil seal locating snap ring. Lift housing and bearings off arm assembly. Slide bearings and spacer out of housing. Remove outer oil seal and snap ring from housing.

b. CLEANING. Wash all parts thoroughly in dry cleaning solvent and dry with compressed air.

c. INSPECTION.

Note. The specifications given in this paragraph are manufacturer's limits on new parts.

(1) *Track wheel.* Track wheels will be considered serviceable provided they are free from deep cuts, grooves, stock separation, tread-cracking, or other damage likely to cause early failure. Rims which are not severely damaged, cracked, or out-of-round are considered

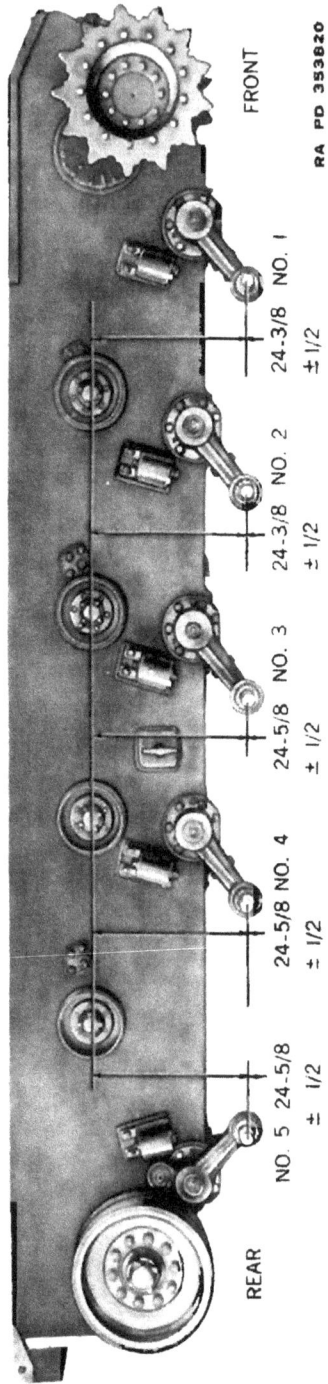

Figure 20. Checking position of support arms.

RA PD 353820

FRONT

NO. 1   24-3/8   ± 1/2

NO. 2   24-3/8   ± 1/2

NO. 3   24-5/8   ± 1/2

NO. 4   24-5/8   ± 1/2

NO. 5   24-5/8   ± 1/2

REAR

34

TRACK SUPPORT ARM AND
HOUSING ASS'Y

RA PD 331905

Figure 21. Removing track support arm housing nut.

BOLT
WASHER

WHEEL
LOCK
NUT

DIRT SHEDDER

INNER CONE

COTTER PIN

GASKET

NUT

HUB CAP

KEYED WASHER

OUTER CONE

OIL SEAL

BEARING SPACER

SEAL SPACER

HUB ASSEMBLY

WHEEL ASSEMBLY

RA PD 353614

Figure 22. Wheel hubs, bearings, and seals, disassembled.

serviceable for both overseas and domestic use. Diameter of wheel mounting holes should be 0.766 to 0.776 inch. Diameter of wheel mounting studs should be 0.742 to 0.750 inch. Fit of wheel on stud should be 0.016 to 0.034 inch (loose).

(2) *Track wheel hub.* Inspect track wheel hub for breaks or cracks. Inspect for stripped studs. Diameter of outer cup seat should be 3.6695 to 3.6710 inches. Outside diameter of cup should be 3.6718 to 3.6728 inches. Fit of cup in seat should be 0.0008 to 0.0033 inch

(tight). Diameter of inner cup seat should be 4.4345 to 4.4365 inches. Outside diameter of inner cup should be 4.4375 to 4.4385 inches. Fit of cup in seat should be 0.001 to 0.004 inch (tight).

(3) *Wheel spindle.* Measure spindle (fig. 23) on support arm for inner cone. Outside diameter of spindle for inner cone should be 2.4993 to 2.4998 inches. Inside diameter of inner cone should be 2.5000 to 2.5005 inches. Clearance between cone and spindle should be 0.0002 to 0.0012 inch (loose). Diameter of spindle for outer cone should be 1.9993 to 1.9998 inches. Inside diameter of outer cone should be 2.000 to 2.0005 inches. Clearance between cone and spindle should be 0.0002 to 0.0012 inch (loose). Inside diameter of slinger should be 5.091 to 5.094 inches. Diameter of slinger seat on hub should be 5.095 to 5.098 inches. Fit of slinger on hub should be 0.001 to 0.007 inch (tight).

(4) *Support arm housing.* Diameter of oil seal seats in housing should be 5.0005 to 5.0020 inches (fig. 23). Outside diameter of oil seals should be 5.002 to 5.006 inches. Fit of oil seal in housing should be 0.0000 to 0.0055 inch (tight). Inside diameter of bearing seats in housing should be 5.0005 to 5.0020 inches. Outside diameter of roller bearings should be 4.999 to 5.000 inches. Clearance between bearing and housing should be 0.0005 to 0.0030 inch (loose). Width of bearing retaining snap ring grooves in housing should be 0.155 to 0.160 inch. Thickness of snap ring should be 0.154 to 0.156 inch. Clearance between ring and groove should be 0.001 to 0.006 inch (loose).

(5) *Support arm.* Measure diameter of support arm shaft at several points along shaft. Diameter of shaft should be 3.4977 to 3.4987 inches. Inside diameter of roller bearings should be 3.4992 to 3.5000 inches. Clearance between bearing and shaft should be 0.0005 to 0.0023 inch (loose). Inside diameter of outer bearing spacer should be 3.500 to 3.502 inches. Clearance between spacer and shaft should be 0.0013 to 0.0043 inch (loose). Inside diameter of inner bearing spacer should be 3.505 to 3.510 inches. Clearance between spacer and shaft should be 0.0063 to 0.0123 inch (loose). Inspect threads on support arm and arm retaining nut for damage caused by staking.

d. Assembly and Installation. (1) *Assemble support arm and housing* (fig. 23). Install snap ring in outer end of support arm housing. Coat edge of new oil seal with gasket cement and tap seal into housing until seal contacts snap ring.

Caution: Tap on outer edge of seal with soft hammer so seal is not damaged.

Lubricate roller bearings generously with seasonal grade general purpose grease and install outer bearing, bearing spacer, and inner bearing in hub. Install inner oil seal locating snap ring in groove in hub. Coat edges of new seal with gasket cement, and tap seal into hub until seal contacts snap ring.

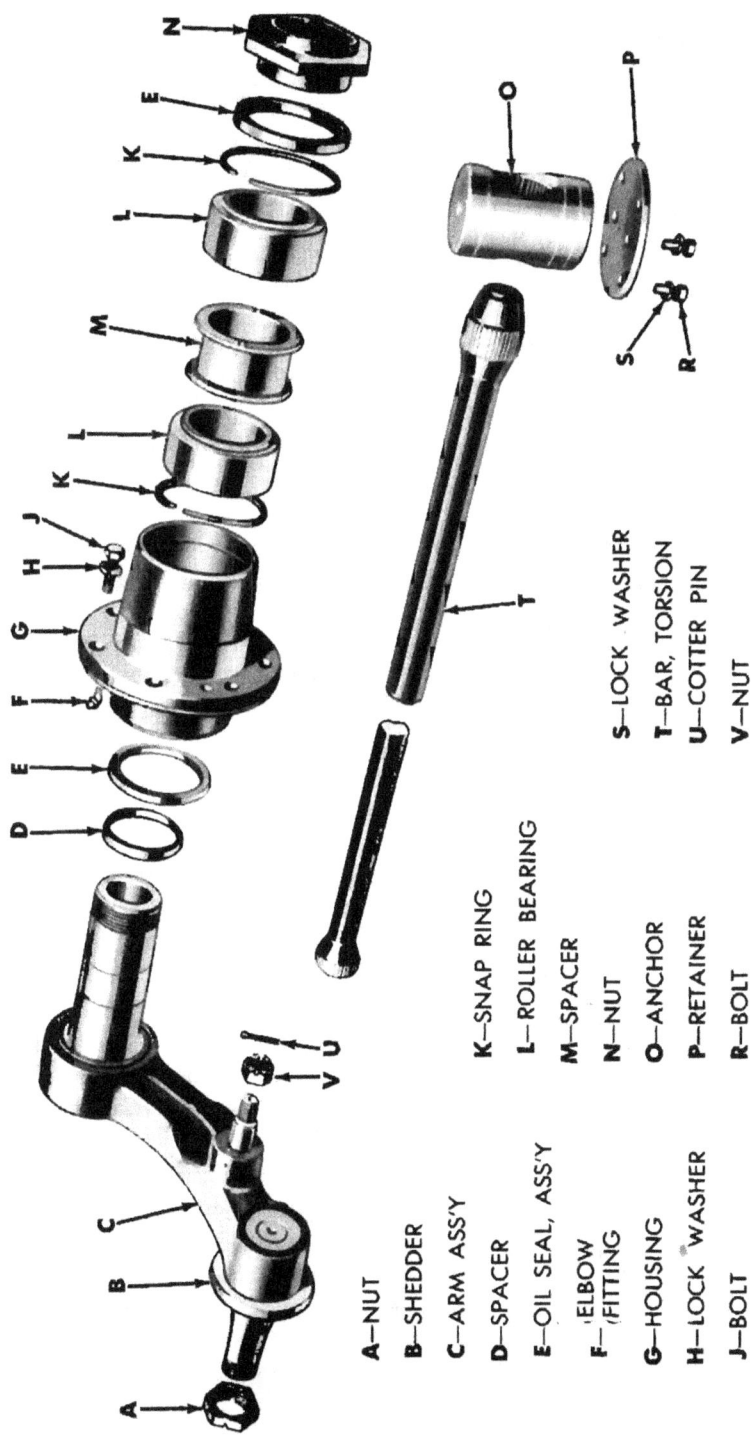

A—NUT
B—SHEDDER
C—ARM ASS'Y
D—SPACER
E—OIL SEAL, ASS'Y
F—ELBOW FITTING
G—HOUSING
H—LOCK WASHER
J—BOLT

K—SNAP RING
L—ROLLER BEARING
M—SPACER
N—NUT
O—ANCHOR
P—RETAINER
R—BOLT

S—LOCK WASHER
T—BAR, TORSION
U—COTTER PIN
V—NUT

RA PD 331906

Figure 23. Support arm, housing, bearings, and torsion bar, disassembled.

Caution: Tap on outer edge of seal with soft hammer so seal is not damaged.

Both seals are to be installed with feather edge towards outside or outer side. Slide spacer over spindle of support arm. Slide hub and bearing assembly over spindle of support arm until outer bearing contacts spacer on spindle. Install large nut holding hub on spindle, using suspension arm bearing nut wrench (41-W-639-395). Tighten nut to 200 foot-pounds minimum. Stake nut after tightening, by bending edge of nut to flat on spindle. Coat support arm housing flange on hull with sealing compound. Lift or hoist assembly into position on hull side, and line up mounting screw holes with drift punch. Dip seven arm-mounting screws in sealing compound, install in housing and tighten to 240-260 foot-pounds.

(2) *Assemble and install hubs and wheels.* Assemble and install wheel hubs and wheels as explained in TM 9-729 or TM 9-744.

19. Torsion Bar

a. REMOVAL. (1) *Remove torsion bars.* Refer to TM 9-729, or TM 9-744. During this operation check support arms and torsion bars (par. 18*a* (2)).

(2) *Remove torsion bar anchors.* Remove screws holding torsion bar anchor cover to bottom of hull, and slide out anchor and cover.

b. CLEANING. Wash all parts thoroughly in dry cleaning solvent and dry with compressed air.

c. INSPECTION.

Note. The specifications given in this paragraph are manufacturer's limits on new parts.

(1) *Torsion bar.* Inspect serrations on both ends of bar (fig. 22) for damage that may cause failure in service. Examine serrations for wear, and check backlash when installed in torsion bar anchor and in suspension arm. Backlash should be within 0.004 to 0.008 inch. Torsion bars must not be repaired by heating or welding because heat destroys the torsion characteristics, and weakens the bars so that they may fail in service. Measure diameter of torsion bar at outer end by placing two 0.1300-inch diameter wires 180° apart in the serrations of the bar. Outside diameter of torsion bar should be 2.442 to 2.444 inches over wire. Measure diameter of torsion bar at anchor end by placing two 0.1300 inch diameter wires 180° apart in serrations of bar. Diameter of torsion bar should be 2.341 to 2.343 inches over wires. Diameter of torsion bars D60591A and D60591B at center should be 1.719 to 1.734 inches. Diameter of torsion bars D60417A and D60417B at center should be 1.563 to 1.568 inches. Diameter of torsion bars 7053474 and 7053475 at center should be 1.875 to 1.890 inches.

(2) *Torsion bar spring test.* Test the torsion bars to determine whether they have been weakened from operation. The angle of twist

and torque foot-pounds are measured by installing bar in vehicle or by using a special fixture, anchoring one end of bar and attaching a 13-inch length lever splined to the opposite end. With hydraulic jack resting on scales, and head of jack under the end of support arm, or of the 13-inch lever, raise the jack to twist torsion bar for obtaining reading. At a 24° angle of twist (normal position in tank), torsion bars D60417A and D60417B read approximately 3,200 foot-pounds. At a 15° angle of twist (normal position in tank), torsion bars D60591A and D60591B read approximately 2,800 foot-pounds. At a 14° 40′ angle of twist (normal position in tank), torsion bars 7053474 and 7053475 read approximately 4,000 foot-pounds.

(3) *Torsion bar anchor.* Inspect serrations in torsion bar anchor for wear. Slide torsion bar anchor over torsion bar anchor end and check for backlash. Backlash should be 0.004 to 0.008 inch. Inspect torsion bar anchor retaining screw holes for stripped threads. Threads can be cleaned by using a 3/8–24NF–2 tap.

d. ASSEMBLY. (1) *Torsion bar arrangement in vehicle.*

M24 Early Series

Position	Part No.
Left No. 1 and 2	D60591B
Right No. 1 and 2	D60591A
Left No. 3 and 4 and right No. 5	D60417B
Right No. 3 and 4 and left No. 5	D60417A

M24 Late Series

Position	Part No.
Left No. 1 and 2	7053474
Right No. 1 and 2	7053475
Left No. 3 and 4 and right No. 5	D60591B
Right No. 3 and 4 and left No. 5	D60591A

M41 Series

Position	Part No.
Left No. 1 and 2	D60591B
Right No. 1 and 2	D60591A
Left No. 3 and 4 and right No. 5	7053475
Right No. 3 and 4 and left No. 5	7053474

(2) *Install torsion bar anchors.* Refer to TM 9–729 or TM 9–744.

(3) *Install torsion bars.* Refer to TM 9–729 or TM 9–744.

20. Track Support Roller, Bearings, and Seals

a. DISASSEMBLY. Remove and disassemble track support roller (TM 9–729 or TM 9–744).

b. CLEANING. Wash all parts thoroughly in dry cleaning solvent and dry with compressed air.

c. INSPECTION.

Note. The specifications given in this paragraph are manufacturer's limits on new parts.

(1) *Track support roller bracket and spindle.* Inspect oil shedder, bearing spacer, and adapter (fig. 24) on the spindle. If inspection indicates damage or wear, remove these parts by pressing off spindle. Inspect bolt holes in mounting bracket for cracks and breaks. Inspect threads on end of spindle for damage. Threads may be cleaned by using a 1¼-12NF-3 die. Check spindle for looseness in bracket, and inspect weld for cracks. Diameter of spindle for inner cone should be 1.6868 to 1.6873 inches. Inside diameter of inner cone should be 1.6880 to 1.6885 inches. Clearance of cone on spindle should be 0.0007 to 0.0017 inch (loose). Outside diameter of spindle for outer cone should be 1.2493 to 1.2498 inches. Inside diameter of outer cone should be 1.2500 to 1.2505 inches. Clearance of cone on spindle should be 0.0002 to 0.0012 inch (loose). Diameter of spindle for adapter should be 1.6895 to 1.6900 inches. Inside diameter of adapter should be 1.6873 to 1.6878 inches. Fit of adapter on spindle should be 0.0017 to 0.0027 inch (tight). Outside diameter of adapter for spacer should be 2.125 to 2.126 inches. Inside diameter of spacer should be 2.129 to 2.139 inches. Clearance of spacer on adapter should be 0.003 to 0.014 inch (loose).

(2) *Track support roller hub and bearings.* Inspect hub for excessive wear and cracks, particularly between the flanges. Inspect all tapped holes for stripped or crossed threads. Examine bearing cups for cracks, chipping, or wear. If hub is serviceable, remove any defective cups and press new ones squarely into hub. Before installing new cups, measure inside diameter of inner cup seat in hub. Diameter of seat should be 3.1225 to 3.1240 inches. Outside diameter of inner cup should be 3.125 to 3.126 inches. Fit of cup in seat should be 0.0010 to 0.0035 inch (tight). Diameter of outer cup seat should be 2.7155 to 2.7165 inches. Outside diameter of outer cup should be 2.717 to 2.718 inches. Fit of cup in seat should be 0.0005 to 0.0025 inch (tight). Diameter of retainer seat in hub should be 3.187 to 3.189 inches. Outside diameter of retainer should be 3.191 to 3.194 inches. Fit of retainer in hub should be 0.002 to 0.007 inch (tight).

(3) *Track support roller disk.* Examine the rubber tire for looseness on the disk, for cracks or pieces broken out, and for wear. If tire is damaged so that it is no longer serviceable, discard the disk assembly. Examine the disk for distortion or out-of-round condition which would cause jumping off the track at high speed. Examine the disk for excessive wear caused by contact with track shoe guides, which weakens the disk and causes failure in service. Discard the disk if it is distorted or worn excessively, as it is not practicable to repair it by

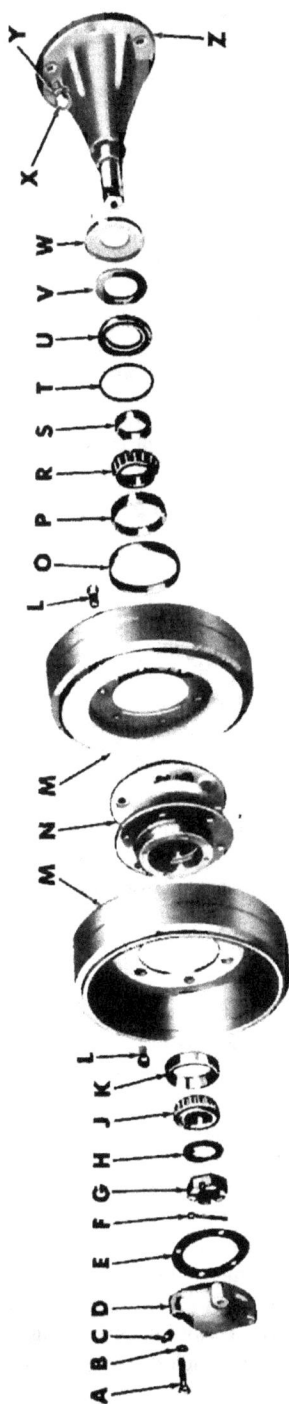

RA PD 353982

A—BOLT
B—LOCK WASHER
C—FITTING
D—CAP
E—GASKET
F—COTTER PIN
G—NUT
H—WASHER

J—CONE
K—CUP
L—BOLT
M—DISK, ASS'Y
N—HUB
O—SLINGER
P—CUP
R—CONE

S—ADAPTER
T—RETAINER
U—OIL SEAL
V—SPACER
W—SHEDDER
X—BOLT
Y—LOCK WASHER
Z—MOUNTING

Figure 24. Track support roller, spindle and bearings, disassembled.

41

bending or welding. Inside diameter of disk mounting holes should be 0.468 to 0.484 inch.

d. ASSEMBLY. Assemble and install track support roller (TM 9-729 or TM 9-744).

21. Suspension Arm Cushion Stop

a. REMOVAL. (1) *Remove volute spring.* Remove bolt and spacing washer that holds volute spring in stop bracket, and remove spring.

(2) *Remove stop bracket.* Remove two bolts inside of stop bracket which hold bracket to hull side wall. Remove two mounting bolts which hold upper end of bracket to hull side (fig. 16) and remove bracket.

Note. No attempt should be made to remove support arm cushion stop brackets which are welded to the hull.

b. CLEANING. Wash all parts thoroughly in dry cleaning solvent and dry with compressed air.

c. INSPECTION. Inspect volute spring for cracks which may cause failure in service. Inspect bracket for cracks, and weld if necessary.

d. ASSEMBLY. (1) *Install bracket.* Identify proper bracket for right and left side. When bracket is properly installed, the boss on the bottom of the bracket points toward the wheel end of support arm. Position stop bracket on hull side wall. Dip two upper screws in sealing compound and install screws and washers. Working under stop bracket, install two lower mounting screws and washers. Tighten all screws to 170–180 foot-pounds.

(2) *Install volute spring.* Position volute spring in stop brackets and enter mounting screw boss on spring in hole in top of stop bracket. Install spacing washer, lock washer, and spring mounting screw, and tighten to 170–180 foot-pounds.

22. Shock Absorbers

a. TESTS BEFORE REPLACEMENT. (1) *Oil leakage inspection.* Slight indications of oil leakage do not warrant shock absorber replacement. A loss of one-half ounce of the type of oil used in these units will completely cover the unit with a film of oil, yet this amount of oil makes no difference in the shock absorber operation. The normal oil capacity of the shock absorber is 32 ounces, but a unit operates satisfactorily with 24 to 26 ounces in reserve cylinder. From the foregoing, it is clear that shock absorbers are never replaced simply for external oil stains. If the lower side of the reserve cylinder (lower part of the unit) is covered with oil, it is probable that the oil level is too low.

(2) *Temperature test.* The temperature method of checking shock absorber operation is practicable only when the following precautions are observed: First, perform this test immediately after a run of not

less than five miles of highway operation, or four miles of cross country operation. The difference in temperature between hull side wall and shock absorber reserve cylinder is then an indication of shock absorber efficiency. If the reserve cylinder has a higher temperature than surrounding parts, shock absorbers are operating satisfactorily. If the reserve cylinder is cold or cylinder temperature is not hotter than surrounding parts, the shock absorber is weak and not operating satisfactorily. The temperature differential should be clearly evident, but the reserve cylinder does not need to be extremely hot to indicate a satisfactory unit.

b. REMOVAL. Remove shock absorbers (TM 9–729 or TM 9–744).

c. DISASSEMBLY OF SHOCK ABSORBER.

Note. When disassembling shock absorbers, it is extremely important that absolute cleanliness be maintained.

(1) *Remove piston and cylinder assembly.* Clean the shock absorber thoroughly before disassembly. Bend the metal tabs away from the windows in the dirt shield, and extend the shock absorber to its fullest extent. If tabs break off, they should be retained, and tack welded in place at time of reassembly. Lay the shock absorber horizontally in a vise, clamping on the outside of the large tube. Unscrew the stop ring, which is engaged in the end of the large tube (fig. 25), using either spanner wrench or drift to loosen. Release the shock absorber from the vise and remount, standing it vertically, and clamping it on the lower forging. Remove the hex head safety screw that holds the locking ring in the reserve tube (fig. 26), then unscrew the locking ring with spanner wrench (41-W-3255-427) (fig. 27). Raise the locking ring, the gasket support washer, and the reserve tube

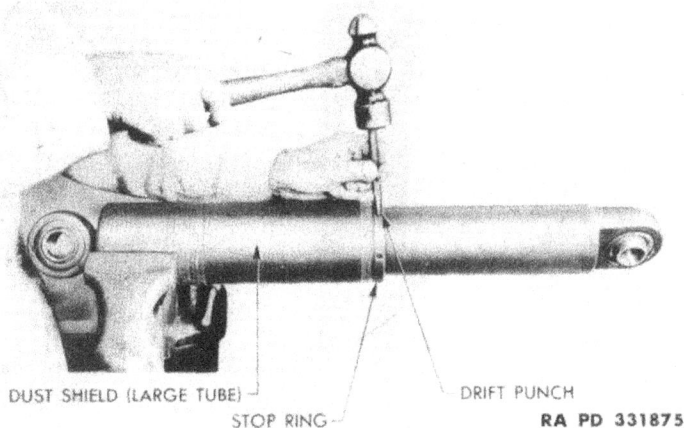

DUST SHIELD (LARGE TUBE)
STOP RING
DRIFT PUNCH
RA PD 331875

Figure 25. Removing shock absorber stop ring.

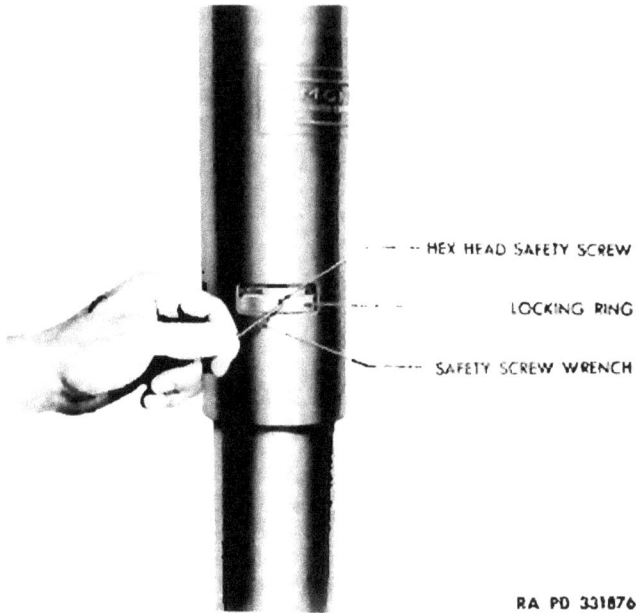

HEX HEAD SAFETY SCREW

LOCKING RING

SAFETY SCREW WRENCH

RA PD 331876

Figure 26. Removing hex head safety screw from locking ring.

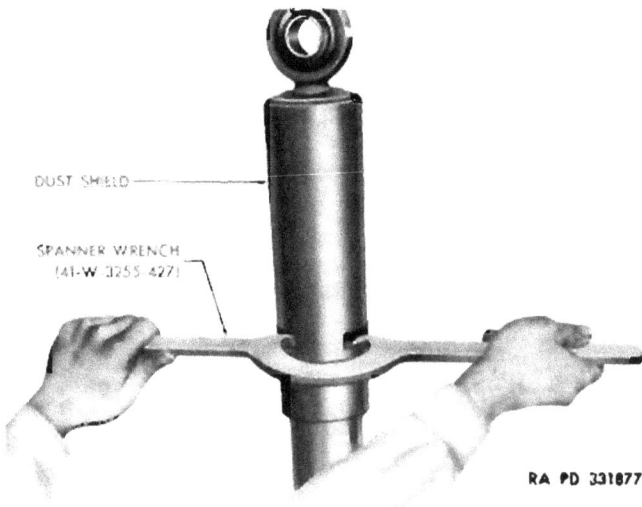

DUST SHIELD

SPANNER WRENCH
(41-W 3255 427)

RA PD 331877

Figure 27. Removing locking ring in reserve tube.

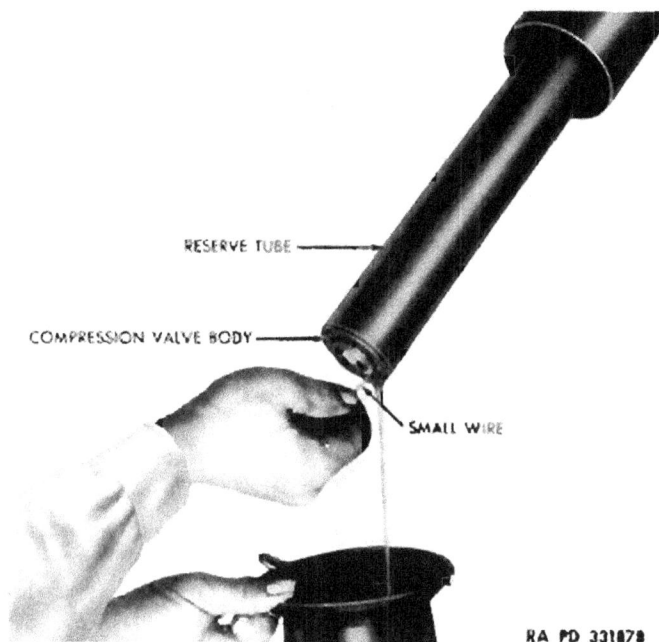

RESERVE TUBE

COMPRESSION VALVE BODY

SMALL WIRE

RA PD 331878

Figure 28. Removing oil from reserve tube.

gasket, pull the piston and cylinder assembly out of the reserve tube, remove the reserve tube from the vise, and pour out the fluid.

(2) *Remove cylinder and compression valve.* Push a small pin through one of the holes in the compression valve body to lift the intake valve off its seat slightly, and drain the fluid out of the cylinder (fig. 28).

Pull the cylinder out of the dust shield until the rod guide and seal assembly is exposed, and carefully tap against the shoulder of the piston rod guide with a brass drift and hammer.

Note. Do not clamp the cylinder in a vise.

When the cylinder and base compression valve have been removed, insert a piece of hard wood through the cylinder and tap the base compression valve out of the opposite end of cylinder (fig. 29).

(3) *Disassemble piston and valve assembly.* Clamp the forged end of the piston rod and dust shield assembly in a vise with the piston up. Remove the nut that holds the piston on the piston rod (fig. 30). Remove the piston rebound valve spring, spring seat, piston and back-up disk, compression valve back-up disk, compression valve spring seat, piston compression valve spring, and guide.

(4) *Remove rod guide and seal assembly.* Place the seal protector

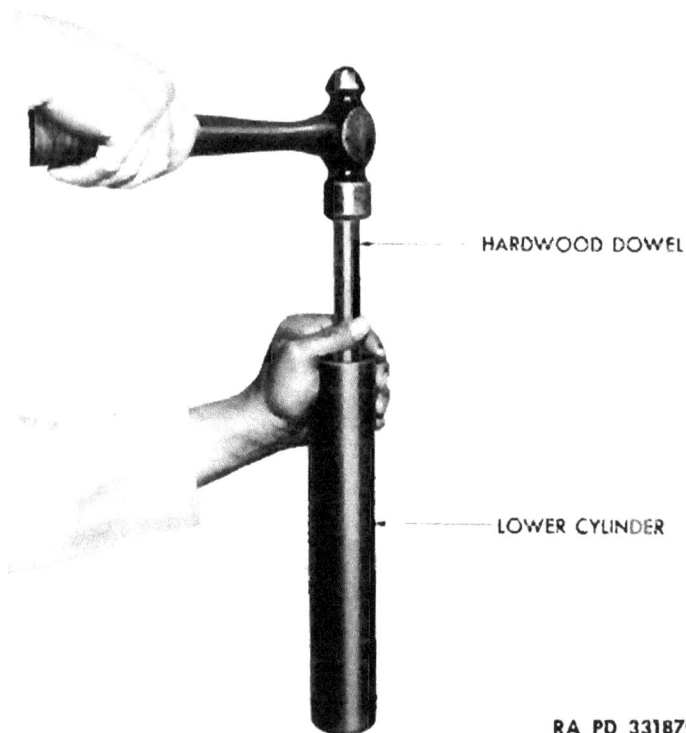

HARDWOOD DOWEL

LOWER CYLINDER

RA PD 331879

Figure 29. Removing base compression valve from cylinder.

pilot (41–P–405) on the end of the piston rod. Hold the pilot firmly in position, and slide the guide and seal assembly off the piston rod. Then remove the pilot. Remove the piston rod and dust shield assembly from the vise, and let the gasket, gasket support washer, and locking ring slide out.

d. INSPECTION. Examine piston rod visually, using a light to make sure that no nicks or scores are present on the surface of the piston rod.

Note. The piston rod and shield assembly are used only if the piston rod is smooth and free from any damage.

Clean the reserve tube thoroughly in dry cleaning solvent, making sure the reserve tube threads are in satisfactory condition. Replace all other parts.

e. ASSEMBLY OF SHOCK ABSORBER. (1) *Install new rod guide and seal assembly on piston rod.* Clamp the dust shield and piston rod assembly in a vertical position in the vise with the open end up. Slide the locking ring on the piston rod with the spanner wrench slots down.

46

PISTON ROD NUT

PISTON REBOUND
VALVE SPRING

SPRING SEAT

PISTON

BACK-UP SPRING

RA PD 331880

Figure 39. Removing piston rod nut.

Slide a new gasket support washer on the piston rod so that the inside diameter is higher than the outside diameter.

Note. It is important that this washer is not assembled upside down. Install a new reserve tube gasket. Insert a length of ½-inch brass rod through the windows of the dust shield to prevent the rod guide and seal assembly from sliding into the dust shield. Place the seal protector pilot (41-P-405) on the piston rod. Slide a new rod guide and seal assembly onto the rod with the seal assembly down (fig. 31). Remove the pilot.

(2) *Assemble new piston and valve assembly on piston rod.* Install the piston compression valve guide, valve spring, spring seat, back-up disk, piston, second back-up disk, second spring seat, piston rebound valve, and piston nut on the piston rod end. Install the lighter of the two valve springs in position first. See that the flange on the valve spring seats protrudes up into the springs and not into the piston. Be sure that the spring seats have been properly entered by the com-

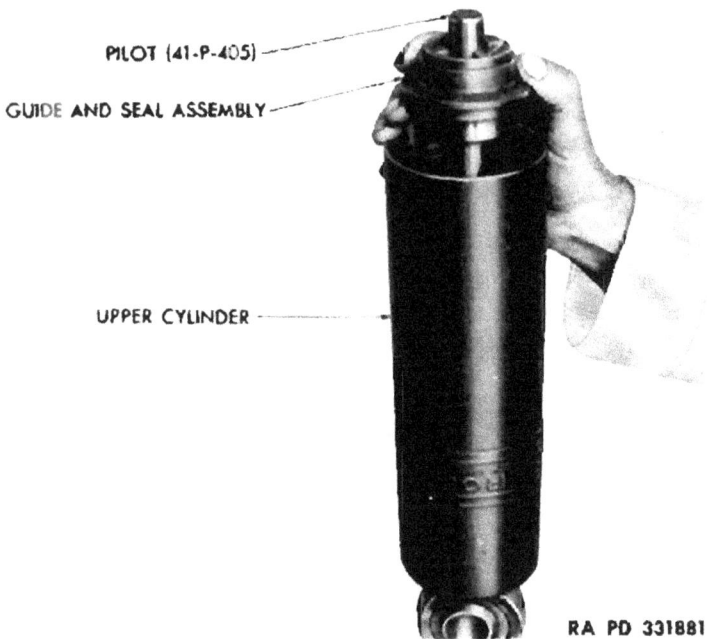

PILOT (41-P-405)

GUIDE AND SEAL ASSEMBLY

UPPER CYLINDER

RA PD 331881

Figure 31. Installing guide and seal assembly on piston rod.

pression valve guide and the piston nut as the nut is being screwed down. Tighten the nut securely to approximately 80 foot-pounds torque. With the brass drift inserted through the dust shield windows, slide a new cylinder over the piston. Place a flat plate over the end of the cylinder, drive the cylinder down, forcing the rod guide and seal assembly squarely into the end of the cylinder and against the shoulder of the rod guide and seal assembly. Remove the brass drift. Slide cylinder, guide, seal assembly up and down along piston and rod.

Note. This assembly must be free from bind. If any bind is noticed, unloosen piston rod enough to rotate piston one-quarter turn. Retighten and repeat check for bind. Repeat this procedure, if necessary, until assembly moves freely.

(3) *Install new fluid and new compression valve assembly.* Pull the cylinder out to its maximum length. Pour 18 ounces of shock absorber fluid (light) into the cylinder (fig. 32). Tap the new compression valve assembly into the end of the cylinder.

(4) *Install piston and cylinder assembly into reserve tube.* Clamp the reserve tube and forging in a vice with the tube in a vertical position. Pour 14 ounces of shock-absorber fluid (light) into the reserve tube. Slide the cylinder into the reserve tube. Position the reserve tube gasket into pocket between the reserve tube and rod guide and

48

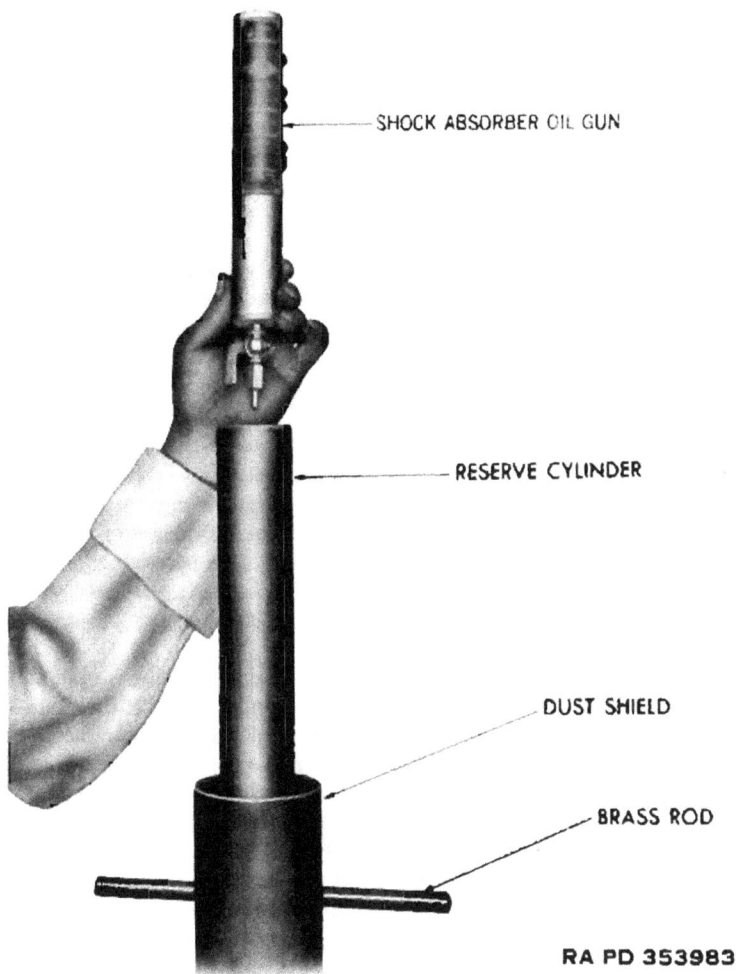

SHOCK ABSORBER OIL GUN

RESERVE CYLINDER

DUST SHIELD

BRASS ROD

RA PD 353983

Figure 32. Refilling shock absorber cylinder.

seal assembly. Fit the gasket support washer against the gasket. Start the threads of the locking ring by hand. Insert the spanner wrench (41-W-3255-427) in the dust shield windows and tighten the locking ring securely (fig. 27). Install and tighten the safety hex set screw. Bend the metal tabs back of the dust shield window, making sure there is no interference with the tube. If the metal tabs were broken off at time of disassembly, tack weld in position. Reassemble the stop ring in the shield. Tighten securely. Prime the shock absorber by reciprocating it two or three strokes.

CHAPTER 6

HULL

Section I. DESCRIPTION AND DATA

23. Description (figs. 7 and 8)

a. The hulls of these vehicles are completely welded structures, except for portions of the front, top, rear, and floor, which are removable for service operations. These removable portions consist of a plate above the controlled differential at the front of the vehicle, two drivers' doors over the drivers' seats, hinged or bolted covers over the engine compartment and radiator compartment, an air inlet and an outlet grille, and removable covers over each fuel tank, and each set of batteries. There are two removable gun crew shields of ¼-inch armor plate around the upper section of the rear compartment of the M41 motor carriage. Openings in the bottom of the hull include the escape door (M24 only), two large inspection plates (one under each engine and transmission), and the small covers just beneath the drain plugs for the engines, hydramatic transmissions, transfer unit, controlled differential, and final drives.

b. The hull floor carries the mounting brackets for the engine and transmission supports, the transfer unit supports, and the differential supports. It also incorporates the tunnels in which the torsion bars for the track suspension wheels are carried. The front of the hull slopes downward at the top and upward at the bottom to form a V. The sides of the hull slope inward at the bottom.

c. On the M24 vehicle, the hull is divided into two compartments, the fighting compartment at the front, and an engine compartment at the rear. On the M41 vehicle, the hull is divided into three compartments, the driving compartment at the front, an engine compartment at the center, and the stowage compartment and gun mount platform at the rear. These compartments are separated by bulkheads that extend from side to side. The bulkhead in front of the engine compartment extends from the roof down to the bulkhead extensions, which in turn extend forward to cover the transfer unit.

d. Seats for the driver and assistant driver are mounted in the front of the hull. These seats have both an up-and-down and fore-and-aft adjustment. Protective pads for driver and assistant driver are provided around the final drive propeller shafts, at the sides of the controlled differential, and on the periscope head rests.

e. The various stowage items carried in the hull are mounted in sheet metal containers, which are bolted or latched to the hull floor and side walls. On M24 vehicles ammunition stowage is provided under

50

hinged covers, which, when closed, serve as a subfloor (fig. 12). The 75-mm. shells are carried in double-walled containers, which are filled with liquid. On M41 vehicles, stowage for 22 rounds of 155-mm. howitzer ammunition is provided in racks built into the forward end of the rear compartment below the gun mount.

f. The M41 vehicles are equipped with a spade assembly, hinged to brackets welded to the rear of hull, and a tailgate, which acts as the gunners' platform when lowered with the spade assembly.

24. Data

Armor plate thickness.	M24 (inches)	M41 (inches)
Upper front plate, at 30° from horizontal	1	½
Lower front plate, at 45° from horizontal	1	½
Hull sides, at front, at 78° from horizontal	1	½
Hull sides, at rear, at 78° from horizontal	¾	½
Hull rear	¾	½
Hull roof	½	½, ⅜
Hull floor, at front	½	½
Hull floor, at rear	⅜	⅜
Engine compartment door (Grille)	⅜	

Section II. OVERHAUL OF SUBASSEMBLIES

25. Drivers' Doors

a. REMOVAL. Remove drivers' doors (TM 9-729 or TM 9-744).

b. DISASSEMBLY (fig. 33). (1) *Remove periscope holder.* Remove eight screws and washers holding periscope ring to driver's door assembly. Lift out ring and gaskets, and slide periscope holder out of door.

(2) *Remove hinge spring shield.* Remove cotter pin and snap spring holding shield to top of door, and lift off shield.

(3) *Remove hinge base.* Remove two screws holding hinge cover to bottom of hinge base. Remove cover. Remove snap ring holding base bearing and base on hinge yoke. Tap on edge of base with soft hammer and drive bearing and base off yoke. Remove snap ring holding bearing in base. Tap on outer race of bearing to remove from base.

(4) *Remove yoke assembly.* Remove driver's door hinge pin by driving on end of pin with a brass drift. Remove key from hinge pin. File one end of rivet holding link to yoke assembly. Remove rivet using a small drift punch. Separate yoke from door and link.

(5) *Remove drivers' door hinge bushings.* Remove both drivers' door hinge rubber-lined bushings from door by driving bushing out of door, using a hammer and brass drift.

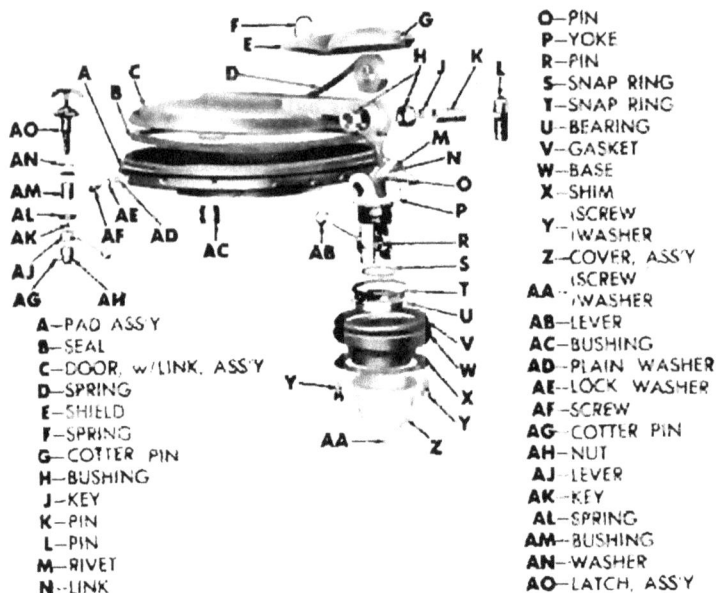

O—PIN
P—YOKE
R—PIN
S—SNAP RING
T—SNAP RING
U—BEARING
V—GASKET
W—BASE
X—SHIM
Y—SCREW
   —WASHER
Z—COVER, ASS'Y
AA—SCREW
    —WASHER
AB—LEVER
AC—BUSHING
AD—PLAIN WASHER
AE—LOCK WASHER
AF—SCREW
AG—COTTER PIN
AH—NUT
AJ—LEVER
AK—KEY
AL—SPRING
AM—BUSHING
AN—WASHER
AO—LATCH, ASS'Y

A—PAD ASS'Y
B—SEAL
C—DOOR, w/LINK, ASS'Y
D—SPRING
E—SHIELD
F—SPRING
G—COTTER PIN
H—BUSHING
J—KEY
K—PIN
L—PIN
M—RIVET
N—LINK

RA PD 331908

Figure 33. Driver's door, disassembled.

c. CLEANING. Wash all parts, except rubber-lined bushings, in dry cleaning solvent, and dry with compressed air.

d. INSPECTION.

Note. The specifications given are manufacturer's limits on new parts.

(1) *Drivers' doors.* Inspect driver's door for cracks or breaks, especially around hinge arm. Diameter of hinge bushing opening should be 1.840 to 1.842 inches. Outside diameter of hinge bushing should be 1.844 to 1.845 inches. Fit of bushing in door should be 0.002 to 0.005 inch (tight). Inside diameter of hinge bushing should be 1.250 to 1.253 inches. Diameter of hinge pin should be 1.246 to 1.248 inches. Clearness between hinge pin and bushing should be 0.002 to 0.007 inch.

(2) *Door hinge yoke.* Check door hinge yoke for breaks or cracks. Check door operating lever to determine if bent. Measure diameter of hinge pin opening in yoke (keyway end). Diameter of this hinge pin opening should be 1.248 to 1.249 inches. Diameter of hinge pin opening (plain end) should be 1.249 to 1.250 inches. Diameter of hinge pin should be 1.246 to 1.248 inches. Fit of pin in yoke should be 0.000 to 0.003 inch (tight). Diameter of bearing seat on yoke should be 3.1494 to 3.1504 inches. Inside diameter of bearings on yoke should be 3.1490 to 3.1496 inches. Clearance of bearing on yoke should be 0.0002 to 0.0014 inch. Width of ring groove in yoke should

52

be 0.162 to 0.172 inch. Width of external ring should be 0.154 to 0.158 inch. Clearance between ring and groove should be 0.004 to 0.018 inch.

(3) *Door hinge base.* Diameter of bearing bore should be 4.9211 to 4.9223 inches. Outside diameter of bearing should be 4.9205 to 4.9213 inches. Clearance between bearing and base should be 0.0002 to 0.0018 inch. Width of ring groove should be 0.162 to 0.172 inch. Width of internal ring should be 0.154 to 0.158 inch. Clearance of ring in groove should be 0.004 to 0.018 inch.

e. ASSEMBLY. (1) *Install drivers' door hinge bushings.* Place driver's door hinge bushing in door opening, and drive or press bushing until bushing is flush with edge of door.

Caution: Tap on outer edge of bushing with soft hammer so bushing is not damaged.

(2) *Install yoke assembly.* Position the two links over the mounting hole of the driver's door and install a new rivet. Peen end of rivet enough to retain links, but not enough to cause binding of link on door. Place driver's door hinge spring in slot in top of door. Slide yoke over link until hinge pin holes in yoke and door line up. Insert hinge pin through first bushing and revolve pin until slot in pin lines up with tang in spring. Tap pin through second bushing. Line up key slot and drive in key, locking hinge pin in position. Connect link to door operating lever by installing clevis pin and a new cotter pin.

(3) *Install hinge base.* Press driver's door hinge bearing into base and install bearing retaining snap ring. Slide base on yoke, and tap into position until bearing retaining ring groove appears on yoke. Install snap ring that holds base on yoke. Slide door hinge cover assembly over operating lever, and install two cover retaining screws.

(4) *Install hinge spring shield.* Place hinge pin shield over hinge spring, and install cotter pin and spring holding cover to door.

(5) *Install periscope holder.* Place periscope holder in opening in top of driver's door. Turn door upside down and install gaskets, ring, and screws holding periscope holder to door. Periscope holder should rotate freely through 360° of travel.

f. INSTALLATION. Install drivers' doors (TM 9-729 or TM 9-744).

26. Ventilator Assembly

a. REMOVAL. Disconnect feed conduit from emergency ignition switch to ventilator assembly, by loosening knurled nut and pulling conduit out of ventilator. Remove four screws holding ventilator assembly to hull roof and remove ventilator.

b. INSPECTION. If ventilator assembly becomes inoperative and resetting the circuit breaker does not correct the condition, remove the

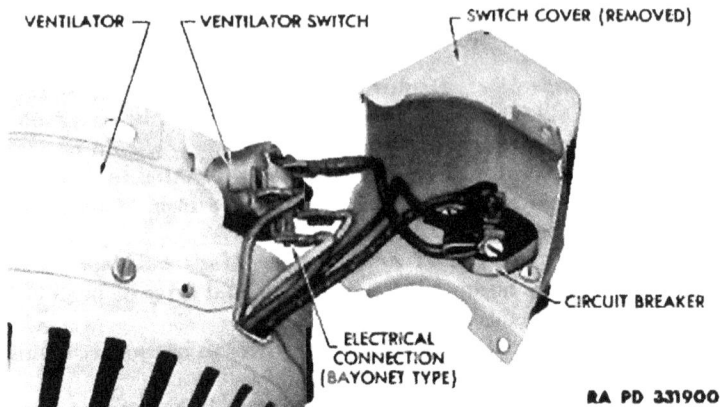

VENTILATOR — — VENTILATOR SWITCH — SWITCH COVER (REMOVED)

CIRCUIT BREAKER

ELECTRICAL
CONNECTION
(BAYONET TYPE)

RA PD 331900

Figure 34. Electrical connections at ventilator switch.

three screws holding the circuit breaker cover to ventilator assembly, and examine wires at switch (fig. 34). These wires have bayonet type terminals and may become disconnected from the switch through vibration. Connect wires to switch in the following order: The black wire to the top center terminal; the red wire to the right terminal (clockwise from the black wire); the yellow wire to the lower terminal; and the brown wire to the left terminal. With the exception of the circuit breaker, there are no spare parts furnished for the ventilator assembly. If assembly is inoperative, replace complete unit.

c. INSTALLATION. Raise ventilator assembly into position over the mounting blocks of the hull roof. Install the four attaching screws and lock washers and tighten securely. Insert the conduit feed cable from the emergency ignition switch into the connection at the front of the ventilator, and tighten the knurled nut securely.

27. Emergency Ignition Switch Box

a. REMOVAL. (1) *Disconnect conduit and right dome light.* Remove two screws holding lens to right dome light. Remove two screws holding dome light to hull roof. Remove small screw and washer holding dome light feed wire to center terminal. Remove wire. Remove two cap screws holding dome light conduit to hull roof.

(2) *Remove emergency ignition switch box.* Disconnect ventilator conduit from front of ventilator, by loosening knurled nut and pulling conduit out of ventilator. Remove three conduits from front of switch box. Remove four screws holding emergency ignition switch box to hull roof and remove box.

b. DISASSEMBLY. (1) *Remove dome light.* Remove two screws holding dome light lens to dome light on bottom of emergency igni-

54

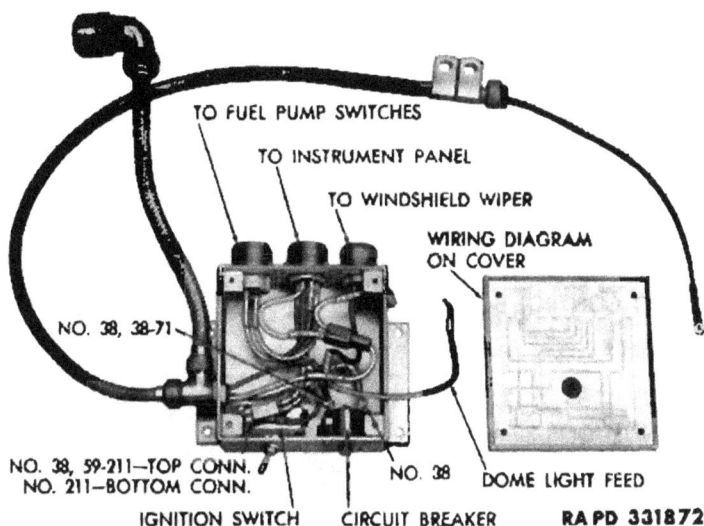

TO FUEL PUMP SWITCHES

TO INSTRUMENT PANEL

TO WINDSHIELD WIPER

WIRING DIAGRAM
ON COVER

NO. 38, 38-71

NO. 38, 59-211—TOP CONN.
NO. 211—BOTTOM CONN.

NO. 38   DOME LIGHT FEED

IGNITION SWITCH   CIRCUIT BREAKER   RA PD 331872

Figure 35. Connections at emergency ignition switch box.

tion switch box. Remove two screws holding dome light assembly
to emergency ignition switch box cover. Remove small screw holding
dome light feed wire to center terminal of dome light. Remove dome
light. Remove four screws holding emergency ignition switch box
cover to box. Lift off cover. Slide dome light wire out of rubber
grommet in cover.

(2) *Remove circuit breaker* (fig. 35). Remove two screws holding
circuit breaker to box. Pull circuit breaker out of inside of box and
remove screws holding terminals to circuit breaker.

(3) *Remove emergency ignition switch* (fig. 35). Remove small
flat nut holding emergency ignition toggle switch to box. Remove
switch from inside of box and remove wires from switch. Note position of wires before removal.

c. ASSEMBLY. (1) *Install ignition switch.* Connect single wire
(No. 211) to bottom terminal of switch. Connect wires (No. 59 and
No. 211) to top terminal of switch. Tighten screws securely. Slide
switch through opening of box and install nut locking switch in box.

(2) *Install circuit breaker.* Connect the single wire (No. 38) to
the outside terminal of the circuit breaker and install the locking
screw. Install the double wire (No. 38 and No. 71) to the other
terminal. Install locking screw. Make sure gasket is in position on
circuit breaker and slide through opening of box. Install two screws
locking circuit breaker to emergency ignition switch box. Slide dome
light feed conduit through rubber grommet in emergency ignition
switch box cover, and install cover.

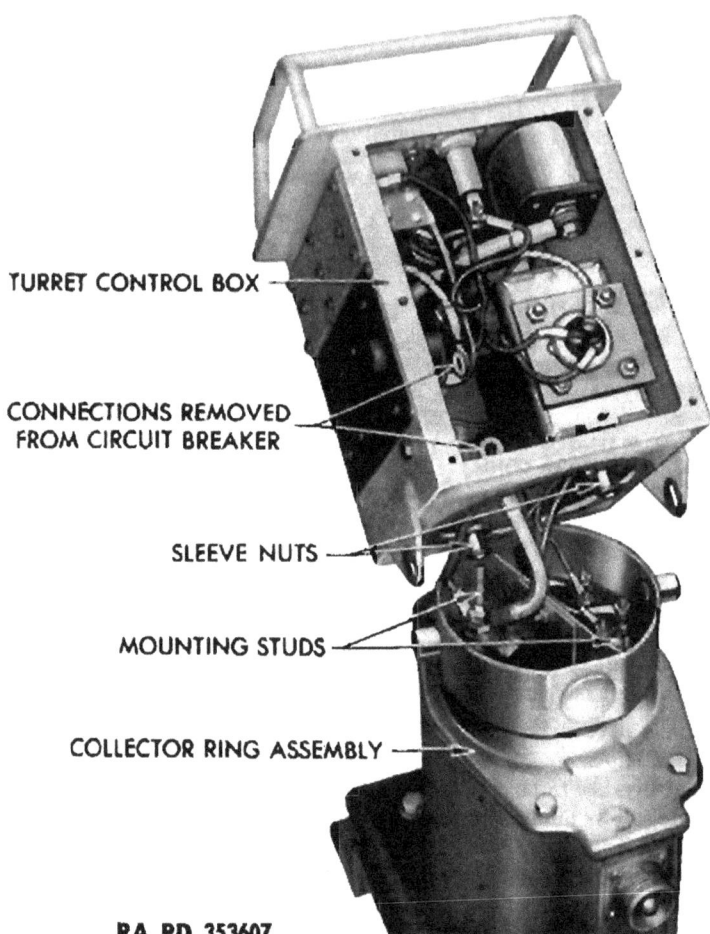

TURRET CONTROL BOX

CONNECTIONS REMOVED
FROM CIRCUIT BREAKER

SLEEVE NUTS

MOUNTING STUDS

COLLECTOR RING ASSEMBLY

RA PD 353607

Figure 36. Disconnecting turret control box from collector ring assembly (M24).

(3) *Install dome light.* Connect the dome light feed wire to the center terminal of the dome light and tighten screw securely. Position the dome light over the two mounting bosses of the cover and install the two dome light retaining screws. Position dome light lens over dome light, and install the two attaching screws.

d. INSTALLATION. Position the emergency ignition switch box over the tapping plates on hull roof, and install the four screws and lock washers. Connect ventilator feed conduit to ventilator and tighten knurled nut securely. Connect the right dome light wire to the center terminal of the dome light, and install the attaching screw. Position dome light over tapping block on hull roof and install the two attaching screws. Place dome light lens over dome light and install the two attaching screws. Install screws holding dome light cable clips to hull roof. Connect fuel pump conduit, instrument panel conduit, and windshield wiper conduit to front of emergency ignition switch box.

28. Collector Ring (M24 Only)

a. REMOVAL. Remove two cap screws holding conduit guard cover to hull floor. Remove cover. Disconnect feed conduit and conduit to radio interphone terminal box. Remove four nuts and lock washers for periscope box, and four nuts and lock washers for oddment box and remove both boxes. Lift collector ring and turret control box out of vehicle.

b. REMOVE COLLECTOR RING FROM TURRET CONTROL BOX. Remove six screws holding cover to rear of turret control box. Remove cover. Remove two cap screws holding turret control box to bosses on side of grounding ring. Unscrew the two sleeve nuts which hold the turret control box to top of grounding ring (fig. 36). Tilt turret control box to one side, and disconnect the radio interphone connecting wires. The wires can be identified both by color and by identification tab at the terminal end. Carefully note the position of the wires in order

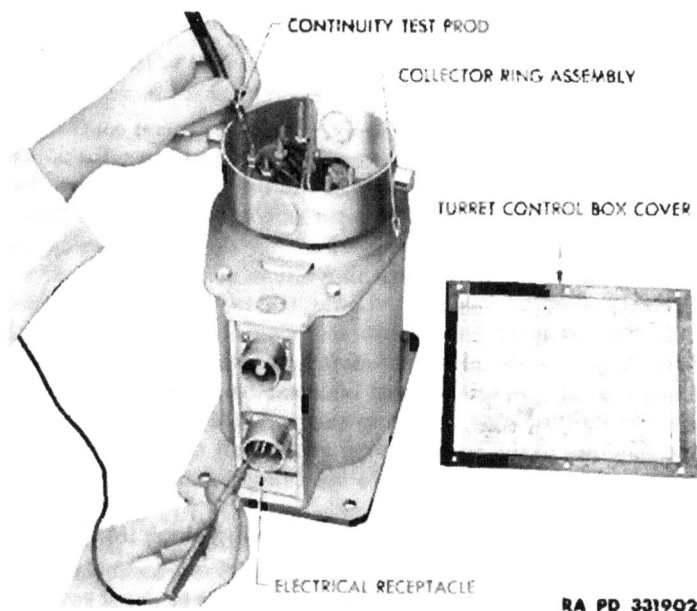

CONTINUITY TEST PROD

COLLECTOR RING ASSEMBLY

TURRET CONTROL BOX COVER

ELECTRICAL RECEPTACLE

RA PD 331902

Figure 37. Checking collector ring electrical circuits (M24).

to reassemble properly. After disconnecting all wires, lay the turret control box to one side.

c. TEST COLLECTOR RING. Test the collector ring to determine if it is serviceable and in good condition. The connections at the radio interphone plug are numbered A, B, C, D, and E. To check each circuit, proceed as follows:

(1) Using a continuity tester (fig. 37), check circuit No. 1 by connecting prods of tester to terminal A and to post No. 1 of the collector ring. Check circuit No. 2 by connecting prods of tester to terminal B and post No. 2 of the collector ring. Check circuit No. 3 by connecting prods of tester to terminal C and post No. 3 of the collector ring. Check circuit No. 4 by connecting prods of tester to terminal D and post No. 4. Check circuit No. 5 by connecting prods of tester to terminal E and post No. 5. If any circuits are found to be open or defective, replace the complete collector ring assembly.

Note. Follow circuit diagram found on inside of turret control box cover.

d. ASSEMBLE COLLECTOR RING TO CONTROL BOX. Position turret control box above collector ring with upper end tilted to one side to allow room for connecting terminal wires to collector ring. Connect the white wire (141) to the No. 1 post of the collector ring. Connect the yellow wire (142) to the No. 2 post. Connect the brown wire (143) to the No. 3 post. Connect the blue wire (144) to the No. 4 post. Connect the black wire (146) to the No. 5 post. Connect the red or wine colored heavy wire (148) to the positive terminal (fig. 36).

Note. Be sure the white cable No. 100 is also connected to the positive terminal. See circuit diagram on inside of control box cover.

Tighten all terminal post nuts securely. Raise the turret control box into a vertical position over top of collector ring and install the two sleeve nuts holding the turret control box to the grounding ring. Install the two cap screws holding the turret control box to the bosses on the side of the grounding ring. Position the back plate on the control box and install the six attaching screws and lock washers.

e. INSTALLATION IN VEHICLE. Lower the turret control box and collector ring assembly into vehicle and place in position on top of the four studs over the propeller shaft housing. Position the periscope spare head stowage box and oddment box over the mounting studs, and install the eight attaching nuts and lock washers. Connect the two conduits to the collector ring and install the conduit guard. Fasten guard securely with the two attaching screws.

29. Gunners' Platform Assembly and Towing Hook (M41 Only.

a. REMOVAL. Lower spade and gunners' platform assembly using spade hoist. Attach a hoist to the tailgate and raise to closed position. Remove the safety pins from the right hinge pin. Drive out the

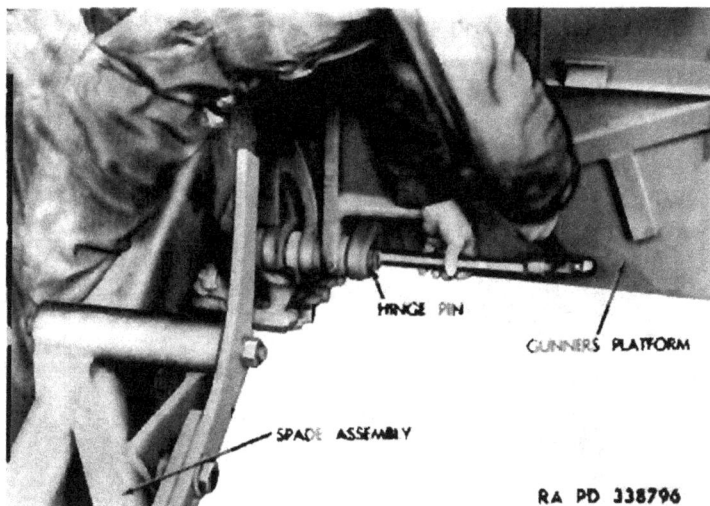

HINGE PIN

GUNNERS PLATFORM

SPADE ASSEMBLY

RA PD 338796

Figure 38. Removing gunners' platform (M41).

hinge pin, releasing towing hook and right hinge of gunners' platform. Repeat this procedure on the left hinge pin. Remove the gunners' platform from the vehicle (fig. 38).

b. CLEANING. Wash all parts in dry cleaning solvent and dry with compressed air.

c. INSPECTION.

Note. The specifications given in this paragraph are manufacturer's limits on new parts.

Inspect the gunners' platform for cracks or breaks, especially at hinge. The inside diameter of the hinge pin hole should be 1.562 to 1.567 inches. The diameter of the hinge pin should be 1.494 to 1.495 inches. Fit of hinge pin in holes should be 0.0670 to 0.0730 inch (loose). Inspect the towing hook for cracks. The inside diameter of the hinge hole in the towing hook should be 1.516 to 1.546 inches.

d. INSTALLATION. Attach a hoist to the gunners' platform and position hinge in brackets on the hull. Drive hinge pins through bracket and hinge from inside. Position towing hook and drive pin through hook and bracket. Install safety pins in hinge pin.

30. Spade Assembly (M41 Only)

a. REMOVAL. Unwind the cable from hoist drum and disconnect by loosening cable V-bolt (fig. 39). Pull the cable from sheaves, and disconnect from spade by removing three cable clamps. Attach hoist to foot of spade to prevent injury to personnel removing spade. Remove lock, bolt, nut, and lock washer from right spade foot (fig. 40).

Figure 39. Disconnecting cable from hoist drum (M41).

Figure 40. Gunner's platform and spade assembly removed from vehicle (M41).

Drive spade hinge pin out. Repeat this procedure on the left side, releasing spade assembly.

b. Inspection and Repair. Inspect the assembly for any broken welds, cracks, or misalinement of components. Weld any cracks or broken welds. See that the cable sheaves rotate freely. If sheave is damaged, weld a new sheave assembly to the spade assembly. Gunner's platform cams must not be bent and must turn freely on bracket. Remove nut, lock washer, cam, spacer washer, bolt and washer to disassemble cam from spade (fig. 41).

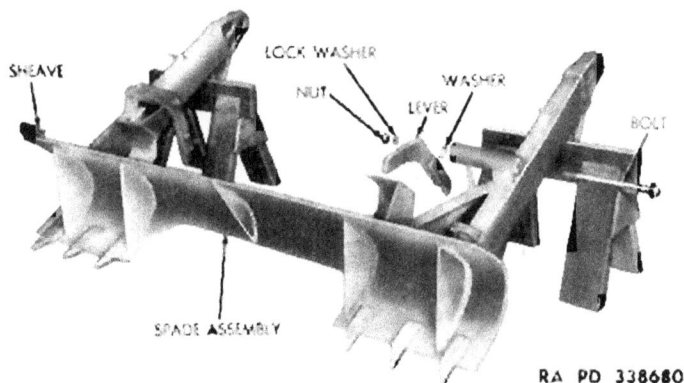

RA PD 338680

Figure 41. Spade assembly removed (M41).

c. Installation. Attach a hoist to the foot of the spade and position in bracket on the rear of vehicle. Drive hinge pin through bracket and hole in spade alining locking hole in spade foot. Slots on each end of pin provide a means of turning pin to aline. Secure with locking bolt, nut, and lock washer. Repeat this procedure on the other spade foot. Attach the cable to the attaching loop on the left side of spade, then pass cable through sheave on left side of hull and down through the sheave on the left corner of the spade. Run the cable through the sheave on the other corner of the spade and leave the free end ready for attaching to the spade winch.

31. Winch Assembly (M41 Only)

a. Description. The winch assembly, located on a platform on the right rear mud guard of vehicle (fig. 10), provides a means of lowering and raising the spade assembly and gunner's platform. The winch is hand-operated from either inside or outside the hull by a crank. A hand-operated brake is used to lower the spade assembly.

b. Removal. The winch cable was removed in paragraph 30. Remove four bolts, nuts, and lock washers from hoist and platform. Remove the hoist from the platform.

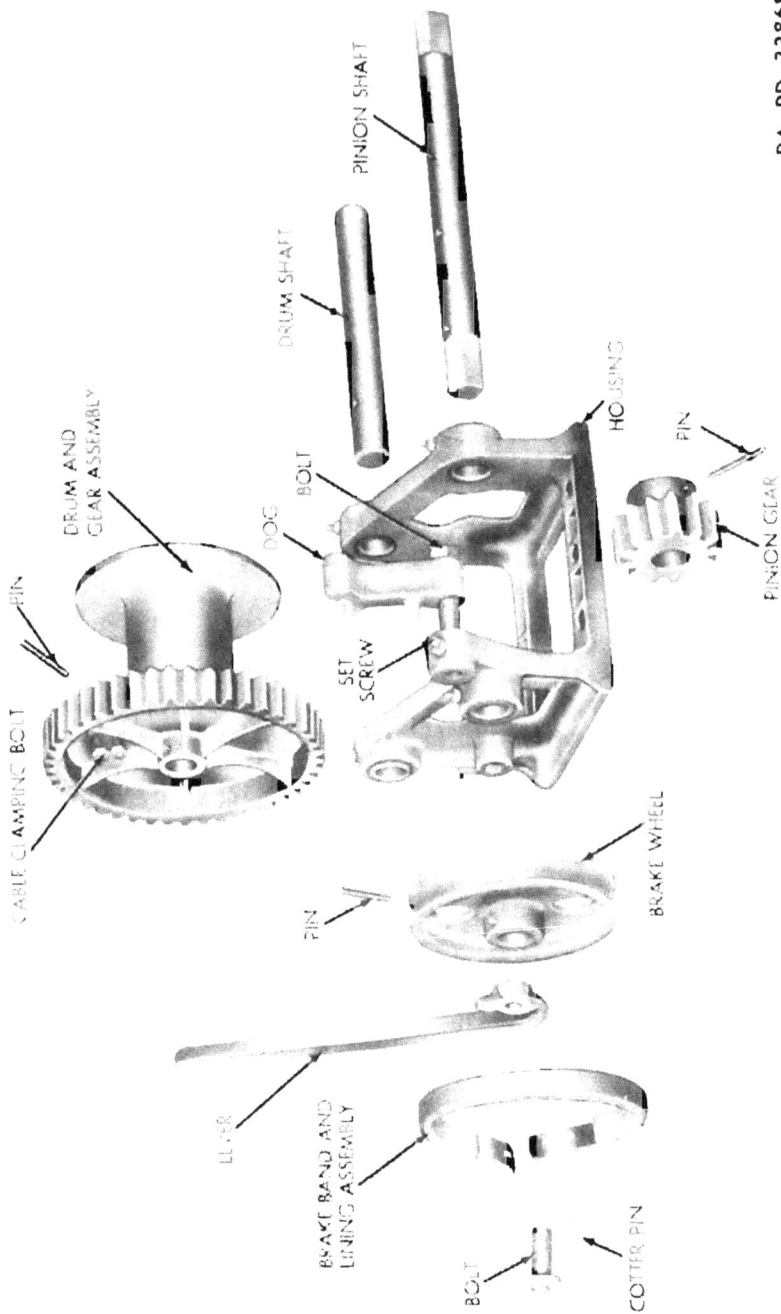

PINION SHAFT

DRUM SHAFT

HOUSING

PIN

PINION GEAR

DRUM AND
GEAR ASSEMBLY

BOLT

DOG

PIN

SET
SCREW

CABLE CLAMPING BOLT

BRAKE WHEEL

PIN

LEVER

BRAKE BAND AND
LINING ASSEMBLY

BOLT

COTTER PIN

RA PD 338681

Figure 42. Winch assembly, exploded view (M41).

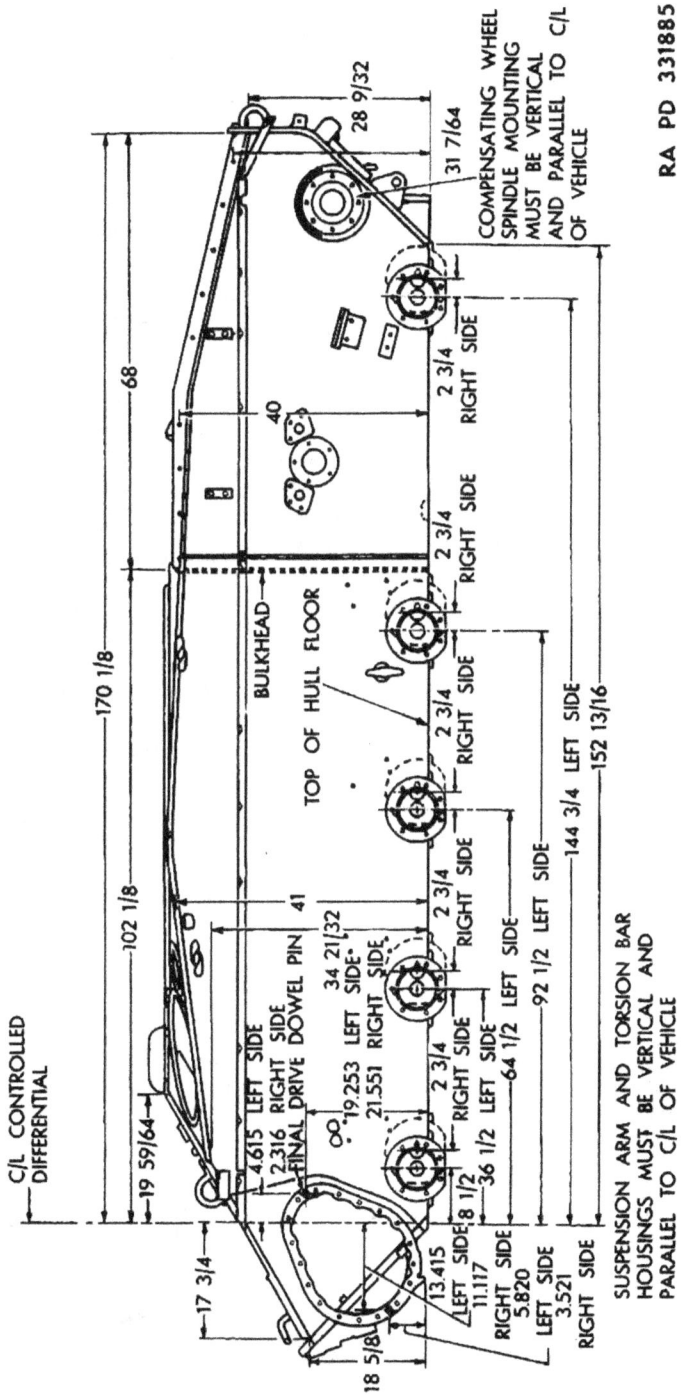

Figure 43. M24 hull dimensions, left side.

RA PD 331885

63

c. DISASSEMBLY (fig. 42). Remove the brake band by removing two cotter pins. Remove nut and lever. Loosen set screw, remove bolt, releasing dog. Drive pins from hub of brake wheel and pinion gear. Drive shaft out of hoist housing, pinion gear, and brake wheel. Drive pin out of cable drum and shaft. Drive shaft from cable drum and housing.

d. CLEANING. Clean all parts in dry cleaning solvent and dry with compressed air.

e. ASSEMBLY. (1) Place housing so that single mounting hole in the base is toward the assembler. Position the drum gear with the gear to the right. Aline pinhole in shaft and drum, and drive shaft into position in drum. Lock with straight pin and peen over the ends of pin.

(2) Start pinion shaft through the housing from the left side. Aline holes in shaft and pinion gear, placing gear on the shaft with gear toward housing (fig. 42). Drive shaft through the gear and housing until pinholes aline. Tap brake drum onto shaft with hub toward outside, keeping pinholes alined. Pin the gear and brake drum to the shaft with two straight pins. Peen over the ends of pins.

(3) Assemble pawl to housing with tooth toward drum gear, tighten bolt, check for clearance, and tighten setscrew (fig. 42).

(4) Assemble brake lever to housing with holes toward brake drum. Tighten nut, check for clearance and stake nut in position. Install brake band and secure to lever with two cotter pins.

f. INSTALLATION. Place the spade hoist on hoist platform on right rear mudguard so that four holes in housing aline with holes in platform. Secure to platform with four bolts, nuts, and lock washers. Slip free end of cable under the cable clamp on drum so that cable comes into drum from the rear of vehicle when the cable clamp is at the bottom of drum. Tighten cable clamp.

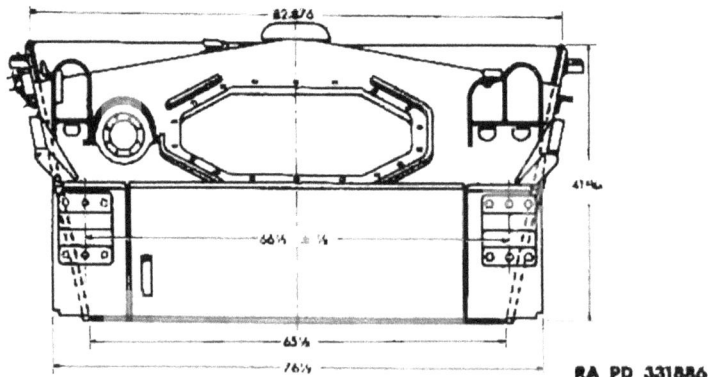

Figure 44. M24 hull dimensions, front.

Section III. ALINEMENT AND REPAIR

32. Scope of Hull Alinement

Alinement of the hull, as covered in this section, consists of checking those surfaces of the hull welded assembly which support the operating units of the vehicle, and in correcting any condition of these surfaces which would cause misalinement of the operating units. The operating units which are affected by distortion of the supporting surfaces of the hull are: Engines, transmissions, differential, final drive assemblies, sprockets, support arm and support arm housings, track wheels, compensating wheels, and track support rollers.

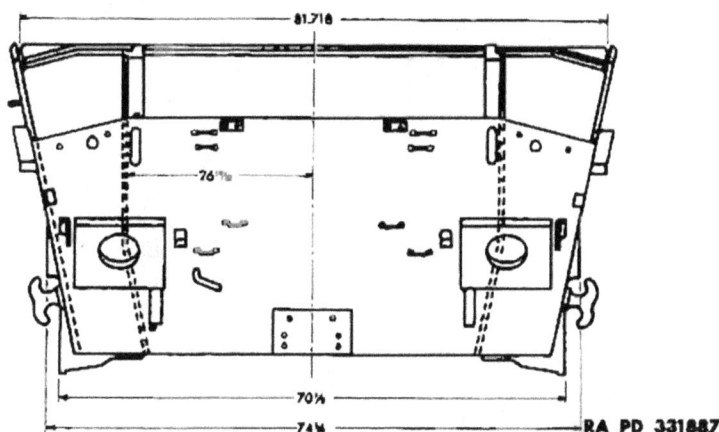

Figure 45. M24 hull dimensions, rear.

33. Preparation of Hull and Visual Inspection

a. PREPARATION OF HULL. The hull must be level, both lengthwise and crosswise, when it is being checked for alinement. Support the hull upon four special stands of equal height, on a level floor. If special stands are not available, other suitable material may be used. If floor is not level, place shims under the stands to make the top end level when checked with a straightedge and spirit level; then place hull on stand. Remove all operating units listed in paragraph 10 so that the supporting surfaces can be inspected and checked.

b. VISUAL INSPECTION. If the vehicle has been in an accident of sufficient force to throw the hull out of alinement, the floor plates and side plates may be distorted, and welded joints cracked or broken. Examine the floor plate and side plate for distortion, and examine all welded joints.

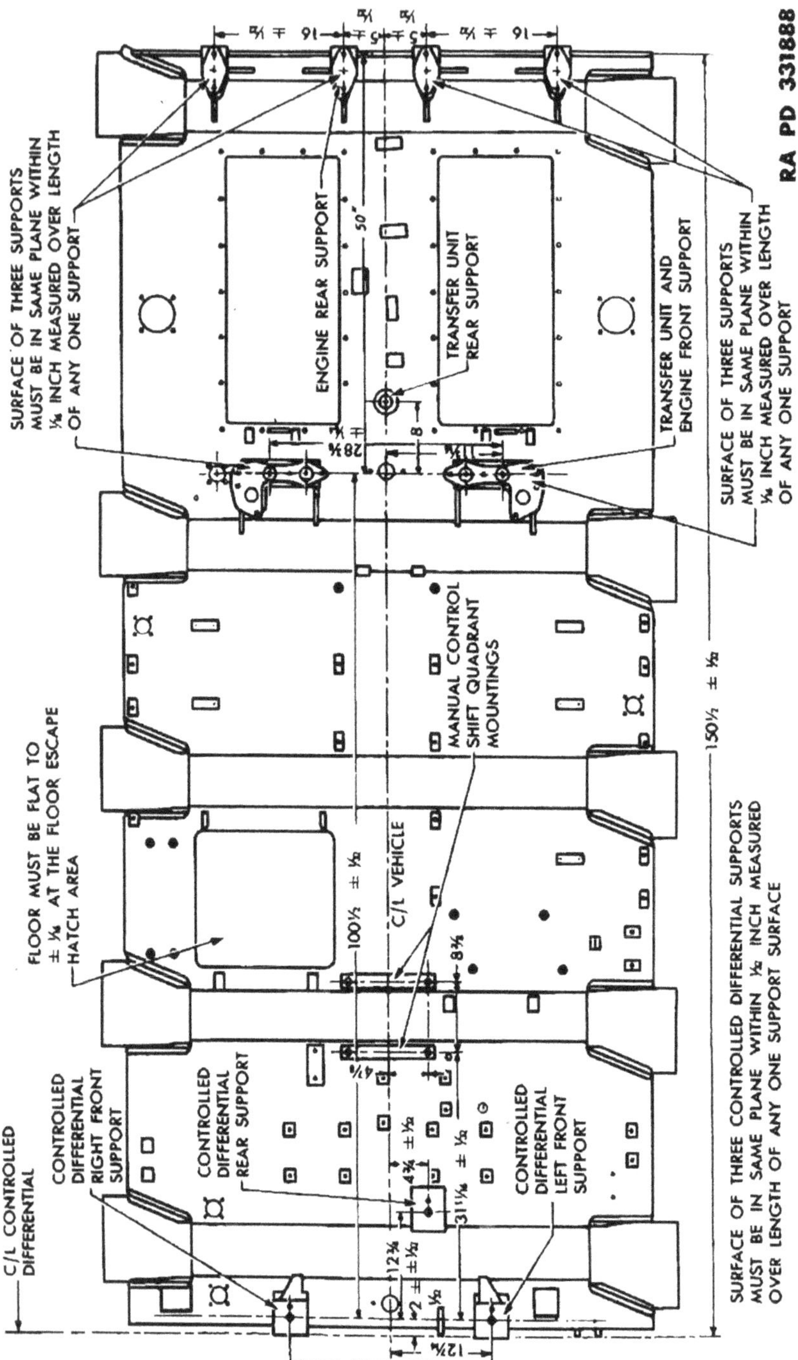

Figure 46. M24 hull dimensions, floor.

RA PD 331888

SURFACE OF THREE SUPPORTS MUST BE IN SAME PLANE WITHIN ⅛ INCH MEASURED OVER LENGTH OF ANY ONE SUPPORT

ENGINE REAR SUPPORT

TRANSFER UNIT REAR SUPPORT

TRANSFER UNIT AND ENGINE FRONT SUPPORT

SURFACE OF THREE SUPPORTS MUST BE IN SAME PLANE WITHIN ⅛ INCH MEASURED OVER LENGTH OF ANY ONE SUPPORT

FLOOR MUST BE FLAT TO ± ⅛ AT THE FLOOR ESCAPE HATCH AREA

MANUAL CONTROL SHIFT QUADRANT MOUNTINGS

C/L VEHICLE

C/L CONTROLLED DIFFERENTIAL

CONTROLLED DIFFERENTIAL RIGHT FRONT SUPPORT

CONTROLLED DIFFERENTIAL REAR SUPPORT

CONTROLLED DIFFERENTIAL LEFT FRONT SUPPORT

SURFACE OF THREE CONTROLLED DIFFERENTIAL SUPPORTS MUST BE IN SAME PLANE WITHIN ½ INCH MEASURED OVER LENGTH OF ANY ONE SUPPORT SURFACE

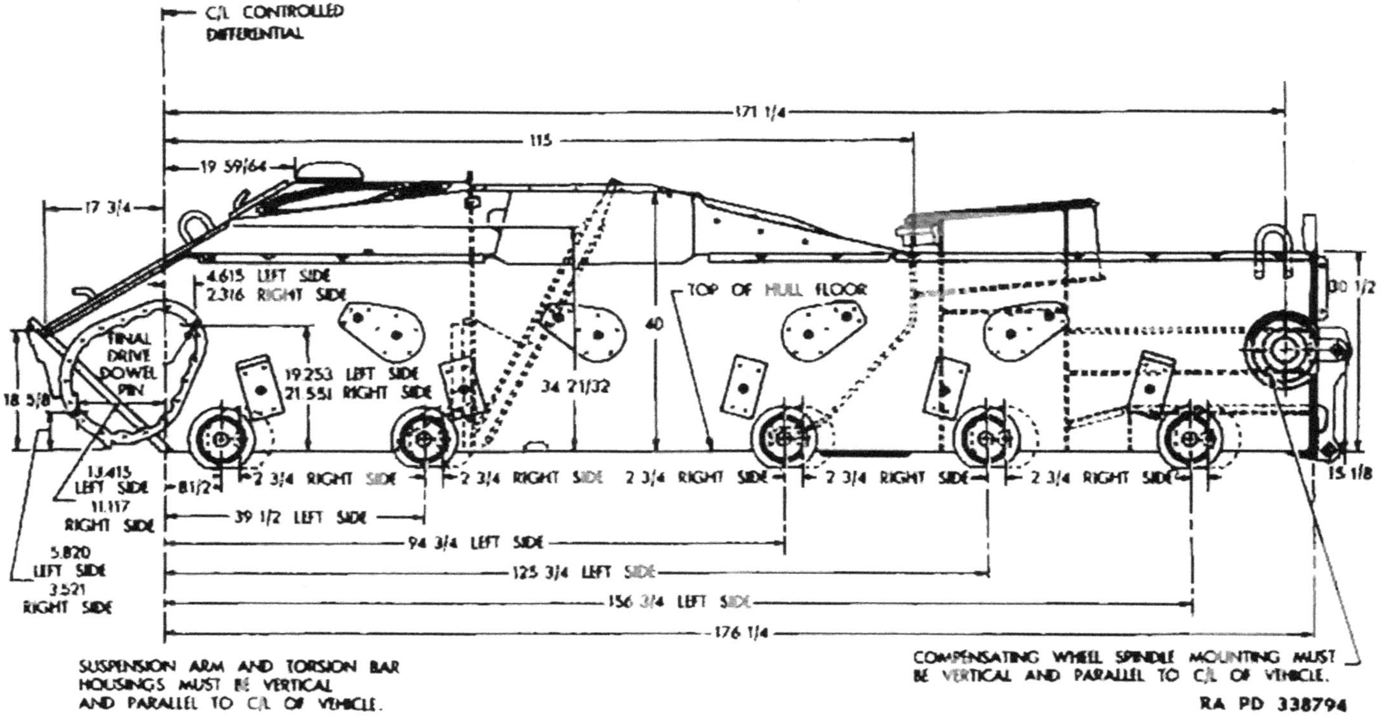

CA CONTROLLED
DIFFERENTIAL

TOP OF HULL FLOOR

FINAL
DRIVE
DOWEL
PIN

171 1/4

115

19 59/64

17 3/4

18 5/8

90 1/2

15 1/8

40

34 21/32

39 1/2 LEFT SIDE
81/2

4.615 LEFT SIDE
2.316 RIGHT SIDE

19.253 LEFT SIDE
21.554 RIGHT SIDE

13.415 LEFT SIDE
11.117 RIGHT SIDE

5.820 LEFT SIDE
3.521 RIGHT SIDE

2 3/4 RIGHT SIDE

2 3/4 RIGHT SIDE

2 3/4 RIGHT SIDE

2 3/4 RIGHT SIDE

2 3/4 RIGHT SIDE

2 3/4 RIGHT SIDE

94 3/4 LEFT SIDE

125 3/4 LEFT SIDE

156 3/4 LEFT SIDE

176 1/4

COMPENSATING WHEEL SPINDLE MOUNTING MUST
BE VERTICAL AND PARALLEL TO C.A. OF VEHICLE.

SUSPENSION ARM AND TORSION BAR
HOUSINGS MUST BE VERTICAL
AND PARALLEL TO C.A. OF VEHICLE.

RA PD 338794

Figure 57. M51 hull dimensions, left side.

67

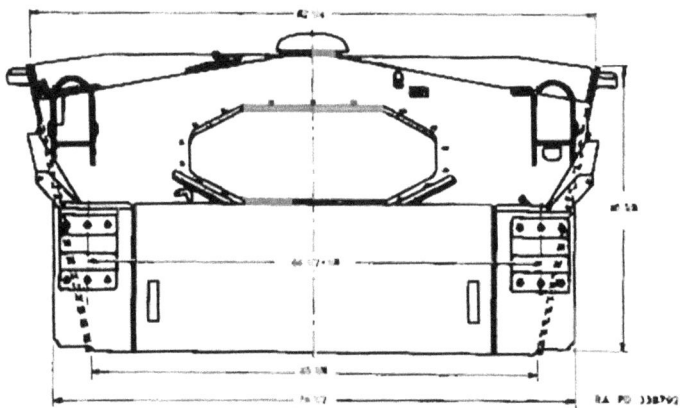

Figure 48. M41 hull dimensions, front.

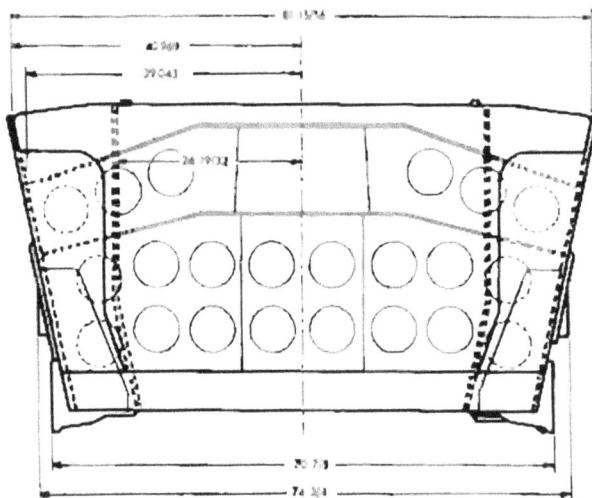

RA PD 338793

Figure 49. M41 hull dimensions, rear.

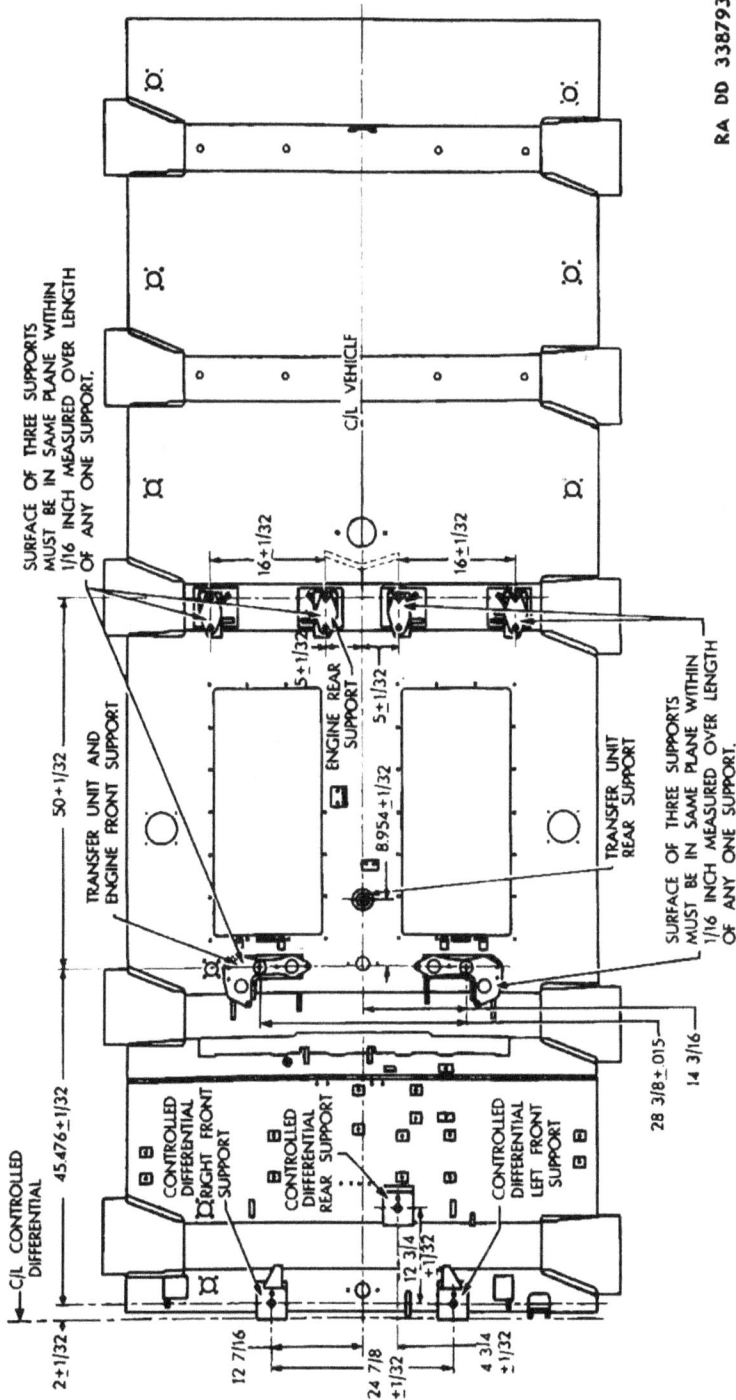

SURFACE OF THREE SUPPORTS
MUST BE IN SAME PLANE WITHIN
1/16 INCH MEASURED OVER LENGTH
OF ANY ONE SUPPORT.

C/L VEHICLE

TRANSFER UNIT AND
ENGINE FRONT SUPPORT

ENGINE REAR
SUPPORT

TRANSFER UNIT
REAR SUPPORT

SURFACE OF THREE SUPPORTS
MUST BE IN SAME PLANE WITHIN
1/16 INCH MEASURED OVER LENGTH
OF ANY ONE SUPPORT.

CONTROLLED
DIFFERENTIAL
RIGHT FRONT
SUPPORT

CONTROLLED
DIFFERENTIAL
REAR SUPPORT

CONTROLLED
DIFFERENTIAL
LEFT FRONT
SUPPORT

C/L CONTROLLED
DIFFERENTIAL

$16 \pm 1/32$

$16 \pm 1/32$

$5 \pm 1/32$

$5 \pm 1/32$

$50 + 1/32$

$8.954 + 1/32$

$45.476 \pm 1/32$

$28\ 3/8 \pm .015$

$14\ 3/16$

$2 \pm 1/32$

$12\ 7/16$

$12\ 3/4 \pm 1/32$

$24\ 7/8 \pm 1/32$

$4\ 3/4 \pm 1/32$

RA DD 338793

Figure 50. M41 hull dimensions, floor.

69

34. Hull Dimensions and Repairs

a. HULL DIMENSIONS. All necessary dimensions for checking the alinement of the M24 hull welded assembly are given in figures 43 through 46. The dimensions for checking the M41 hull are given in figures 47 through 50. After straightening and welding operations, check the hull to make certain that it conforms to the dimensions given in these figures.

b. STRAIGHTENING PLATES. If hull floor plate is bent upward sufficiently to cause misalinement of engine support, transfer unit mounting support, controlled differential, or other hull attaching parts, straighten plate by means of a hydraulic jack and a heavy steel beam placed between hull roof and floor plate. Use care in placing the beam against the roof plate so that the load is evenly distributed to prevent distortion of roof plate. The floor area surrounding the engine and transfer unit mounting supports must be flat within one-sixteenth inch, plus or minus. Examine floor area surrounding the escape door. If floor is bent, straighten it until it is flat within one-sixteenth inch, plus or minus.

c. REPAIRING BROKEN WELDS. Before cracked or broken joints are repaired by welding, make certain that hull is firmly supported in a level position (par. 33) and that hull conforms to dimensions given in figures 43 through 50. If the lifting eyes are damaged, weld new eyes to the hull.

70

CHAPTER 7

TURRET (M24 ONLY)

Section I. DESCRIPTION AND DATA

35. Description

a. The turret used on M24 vehicles is of curved armor plate, all-welded, the inside diameter being approximately 60 inches. The sides of the turret are 1 inch thick; the roof is one-half inch thick. The turret rotates 360° in either direction on a continuous ball bearing mount. This bearing is completely enclosed for protection from direct hits, and from dirt or water. The turret is traversed either by hand or by hydraulic traversing mechanism. An azimuth indicator is installed in the turret to indicate the position of the gun.

b. The turret is circular in shape, except that a bulge extends to the rear (opposite the gun). The radio is mounted in this bulge. A steel box with a hinged cover is bolted on the outside rear end of the bulge to provide space for stowage of tools and spare parts. There is no turret basket, but seats for the commander, gunner, and loader are attached to the turret and rotate with it.

c. Two hinged doors provide access to the turret. One door is on top of the commander's cupola and the other is on the right side of the turret roof, over the loaders seat. The commander has vision through a periscope in the top of the cupola door, and six vision blocks at the base of the cupola. A periscope for the gunner is located forward of the cupola. A pistol port is located on the right side of the turret to permit discarding empty 75-mm. shells with the turret closed.

d. A 75-mm. gun is mounted in the turret, with a separate armor plate casting attached to the gun mount, M64, shielding the gun opening. A caliber .30 machine gun is also mounted in this casting. The caliber .30 machine gun is elevated or lowered manually. The 75-mm. gun is elevated or depressed manually, or with the stabilizer. A caliber .50 antiaircraft machine gun is mounted on a tripod on the turret roof in back of the turret door.

e. A two way ventilator is installed on roof between the two driver's seats to draw air in from outside the tank.

36. Data

Diameter of turret race ----------------- 63 1/2 inches
Number of balls in turret race ----------- 132
Diameter of turret race balls ----------- 1.0000 inches
Weight of turret, less guns and mounts ---- 3,500 pounds
Thickness of turret wall ----------------- 1 inch
Material, turret wall -------------------- Armor plate
Material, turret face plate -------------- Armor steel casting,
welded in place.

Section II. REMOVAL OF TURRET

37. Removal Procedure

a. REMOVE AZIMUTH INDICATOR. Traverse turret to forward position. Remove screws holding azimuth indicator to turret and lift out indicator. Lay indicator carefully to one side to avoid damage.

b. DISCONNECT FEED CABLES. Turn master switch to OFF position. Disconnect three conduits from front of turret control box by loosening knurled nuts and pulling conduits out of box (fig. 51). Disconnect one end of radio ground strap on box.

c. REMOVE TURRET ASSEMBLY. Remove 42 screws holding turret ball ring to hull roof. Turn turret traversing lock to the released position. Attach turret sling (41-S-3832-54) to lifting eyes on turret (fig. 52); remove turret assembly. Place turret in suitable stand.

TURRET CONTROL BOX   ELECTRICAL CONDUITS

RA PD 331909

Figure 51. Disconnecting conduits from control box (M24).

72

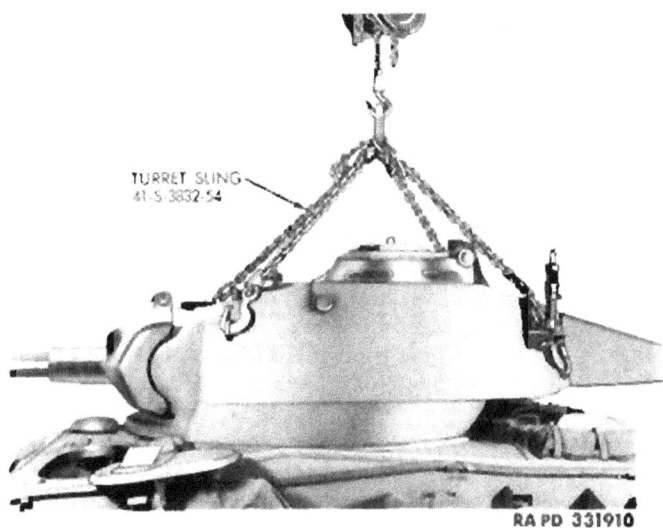

Figure 52. Removing turret assembly (M24).

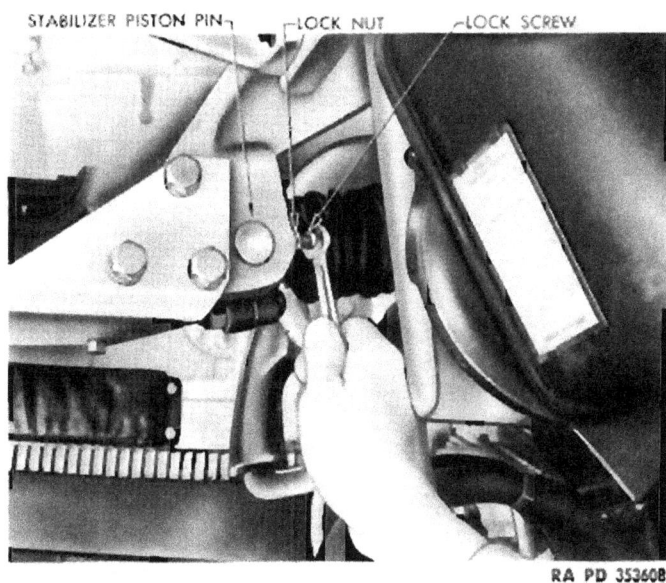

Figure 53. Loosening lock nuts on stabilizer cylinder and piston pin (M24).

Section III. DISASSEMBLY INTO SUBASSEMBLIES

38. Disassembly Procedure

a. DISCONNECT STABILIZER CYLINDER AND PISTON FROM GUN CRADLE. Loosen the two lock nuts (fig. 53) and lock screws and slide pin out of rear end of stabilizer cylinder and piston. Do not disconnect cylinder and piston from bracket on front of turret.

b. REMOVE ELEVATING QUADRANT. Remove three screws holding elevating quadrant and bracket to rear end of stabilizer cylinder and piston bracket.

c. REMOVE ELEVATING MECHANISM. Remove four screws from cover at front end of stabilizer control box. Loosen cable clamp screw and slide cable out of stabilizer control box (fig. 54). Cut locking wire and unscrew turnbuckle from periscope linkage. Remove four screws holding gun elevating mechanism to front of turret (fig. 55). Support assembly while removing last screw.

Caution: Do not lose shims between gun elevating mechanism and front of turret. Tag shims from the upper half and lower half in order to reassemble in the same position.

d. REMOVE TURRET TRAVERSING AND TRAVEL LOCKS. Remove two screws and washers holding turret traversing lock to front of turret

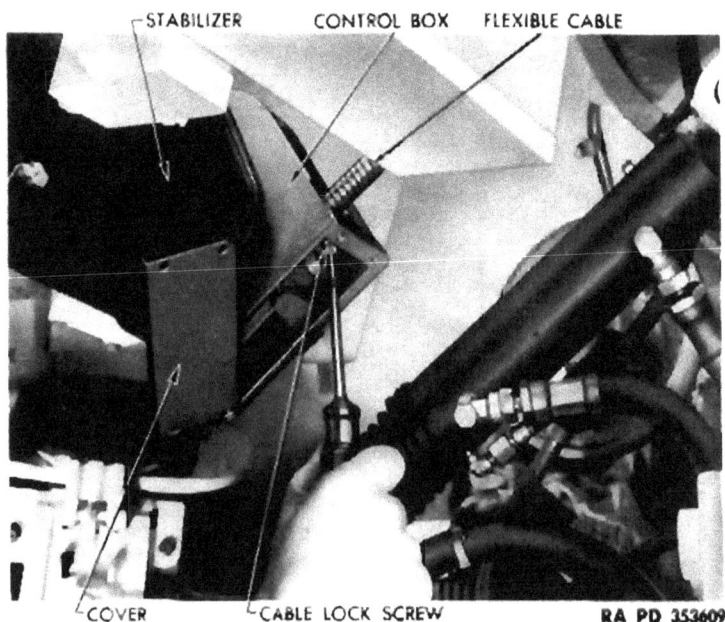

Figure 54. Removing cable from stabilizer control box (M24).

ball ring. Remove lock. Remove cotter pin from travel lock connecting pin. Remove pin and travel lock.

e. REMOVE STABILIZER. Disconnect electrical conduit from stabilizer by pulling the multiprong plug out of stabilizer. Remove four screws holding stabilizer to stabilizer mounting bracket. Remove stabilizer.

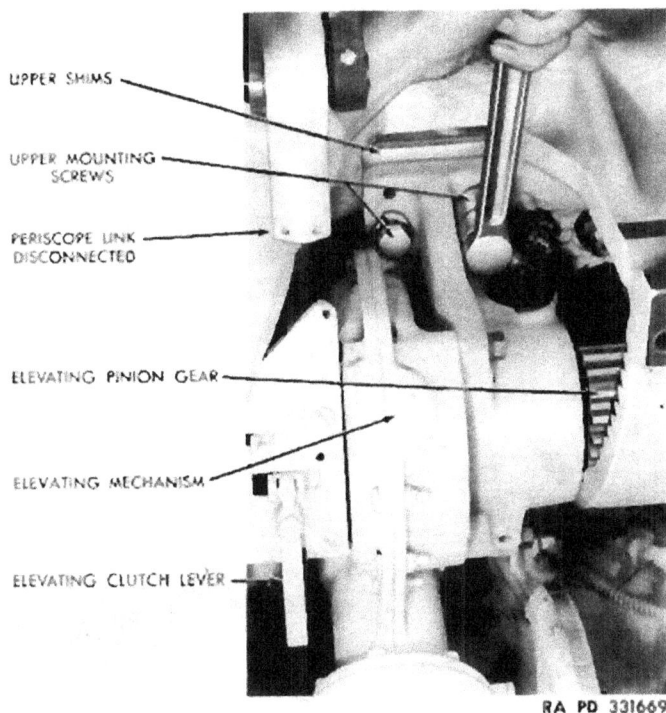

UPPER SHIMS

UPPER MOUNTING SCREWS

PERISCOPE LINK DISCONNECTED

ELEVATING PINION GEAR

ELEVATING MECHANISM

ELEVATING CLUTCH LEVER

RA PD 331669

Figure 55. Removing elevating mechanism (M24).

f. REMOVE STABILIZER MOUNTING BRACKET FROM RECOIL CYLINDER. Remove four screws and washers holding stabilizer mounting bracket to bottom of recoil cylinder. Remove bracket.

g. REMOVE COMMANDER'S AND GUNNER'S BACK REST PADS. Remove screws and washers holding commander's and gunner's back rest pads to turret and remove pads.

h. REMOVE COMMANDER'S AND GUNNER'S SEAT ASSEMBLY. Remove three bolts and nuts holding gunner's seat bracket to turret motor mounting bracket. Remove screws holding commander's seat to rear end of turret. Remove seats and support as an assembly.

i. REMOVE BREECH GUARD. Remove four large nuts and washers holding breech guard to gun cradle. Remove breech guard.

j. REMOVE TELESCOPE MOUNTING BRACKET. Remove screws and washers holding telescope mounting bracket to elevating sector (fig. 56). Lift telescope bracket off dowel pin and remove bracket.

k. REMOVE FIRING MECHANISM. Trip mechanism by pressing hand firing button. Remove three bolts and nuts holding solenoid to firing

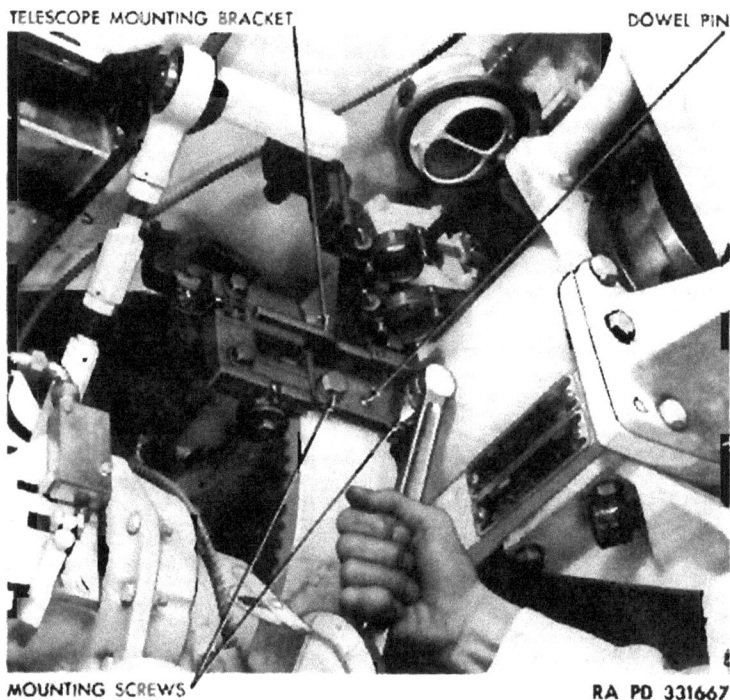

TELESCOPE MOUNTING BRACKET   DOWEL PIN

MOUNTING SCREWS   RA PD 331667

Figure 56. Removing telescope mounting bracket (M24).

mechanism bracket. Remove three screws holding front end of firing mechanism to recoil cylinder (fig. 57). Remove two screws holding firing mechanism to guide and ejector mechanism. Slide recoil switch operating lever out from behind gun retaining nut. Disconnect hand firing plunger from firing mechanism by removing cotter pin and sliding out plunger. Remove four cap screws attaching front end of machine gun bracket to right hand trunnion bearing cap. Remove four cap screws attaching rear of machine gun bracket to side of 75-mm. gun, disconnect solenoid and remove machine gun bracket.

l. REMOVE GUIDE AND EJECTOR MECHANISM. Remove screws and washers holding guide and ejector mechanism to bottom of gun cradle (fig. 58). Slide mechanism off of dowel pins.

m. REMOVE 75-MM. GUN AND GUN MOUNT. Attach a hoist to the lifting eye of the gun shield and take up weight of assembly. Remove nuts holding bearing cap assembly on trunnion. Remove bearing caps. Elevate muzzle and carefully maneuver gun through turret opening to prevent damaging parts. Place assembly to one side and cover with clean cloths to prevent dirt from getting into gun parts.

n. REMOVE GUNNER'S PERISCOPE HOUSING. Remove six nuts and washers from bolts holding gunner's periscope housing to roof of turret. Remove housing and discard gasket.

o. REMOVE STABILIZER CYLINDER AND PISTON, AND BRACKET. Disconnect two rubber flexible tubes (fig. 58), from stabilizer cylinder and piston, and turret motor. Cover openings with masking tape to prevent dirt from entering system. Remove four screws and washers holding cylinder and piston mounting bracket to front of turret and remove cylinder and piston, and bracket as an assembly.

p. REMOVE TURRET MOTOR, OIL PUMPS, AND TRAVERSING MECH-ANISM OIL RESERVOIR ASSEMBLY. Disconnect and remove copper tube from stabilizer oil pump and stabilizer oil reservoir on turret roof (fig. 60). Disconnect and remove one oil tube from front of oil pump to connection on bottom of reservoir.

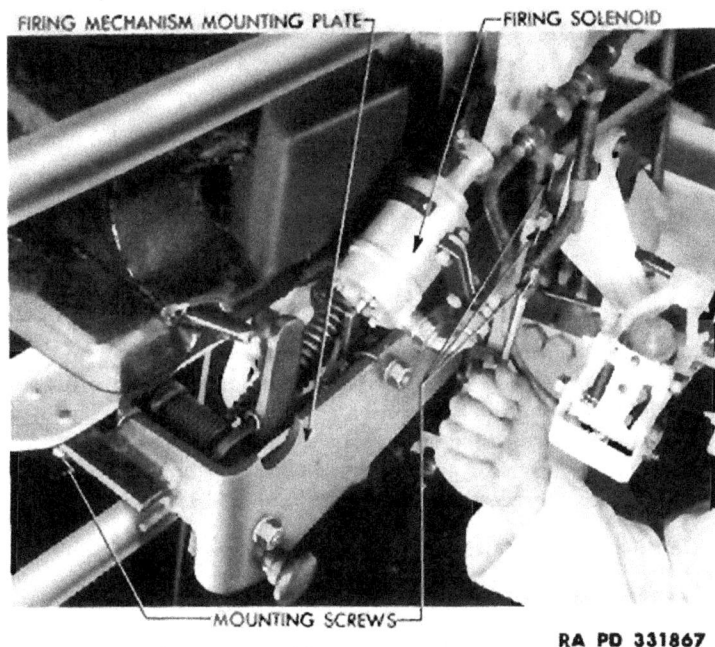

FIRING MECHANISM MOUNTING PLATE— ┌—FIRING SOLENOID

—MOUNTING SCREWS—┘

RA PD 331867

Figure 57. Removing firing mechanism (M24).

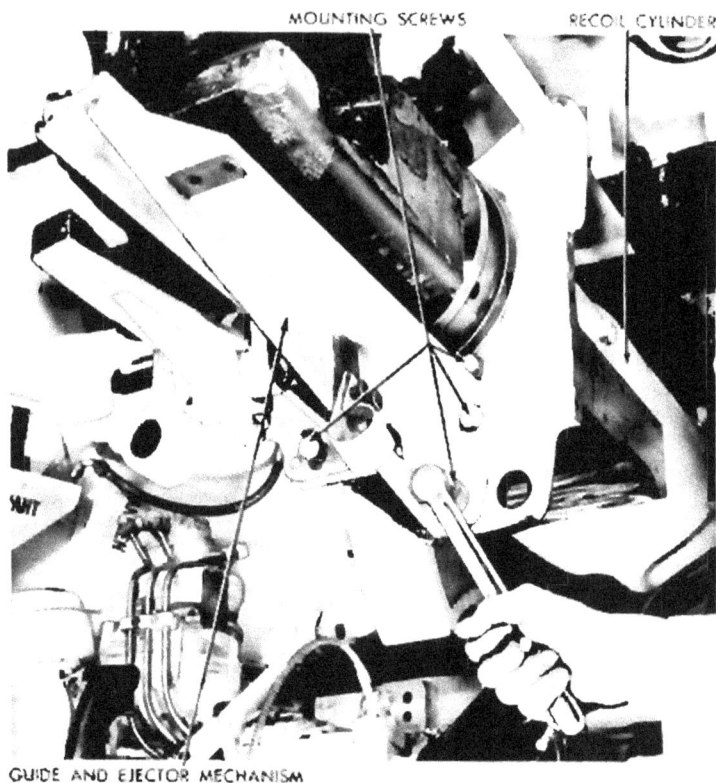

MOUNTING SCREWS   RECOIL CYLINDER

GUIDE AND EJECTOR MECHANISM

RA PD 331871

Figure 58. Removing ejector mechanism (M24).

Caution: Place clean container under reservoir to prevent oil from running out on floor.

Disconnect and remove one oil tube from front of oil pump to connection on top of reservoir. Disconnect and remove two oil tubes from top of oil pump to connections on top of reservoir. Disconnect and remove three oil tubes from oil gear motor to connections on top of reservoir. Cover all openings with masking tape. Remove four screws holding turret motor mounting bracket to turret.

Caution: Support assembly while removing screws.

Remove turret motor, oil pumps, and traversing mechanism oil reservoir as an assembly (fig. 60).

q. REMOVE TRAVERSING GEAR BOX. Remove two screws with 3/4-inch heads holding gear box mounting bracket hinge (fig. 62) to side of turret. Remove lock nut, nut, washer, and spring from antibacklash adjustment stud. Reinstall both nuts and lock together to remove stud. Remove stud. Remove screw with 1-inch head in front of gear box. Remove two screws with ¾-inch heads in front of manual control wheel on top of base ring, and remove traversing gear box assembly.

Caution: Support assembly while removing last screw.

r. REMOVE TURRET RING GEAR FROM TURRET. Remove 16 remaining screws with internal tooth lock washers holding turret to ring gear. Remove 4 nuts from 4 studs on front of turret. Attach a sling to lifting eyes of turret and remove turret from turret ring gear (fig. 63). Remove oil seal retaining spring from outer edge of ring gear. Remove rubber oil seal from outer edge of ring gear. Note that feather edge of seal is toward top of gear.

s. REMOVE ANTIAIRCRAFT TRIPOD. Remove six bolts and nuts holding antiaircraft tripod to top of turret. Remove tripod.

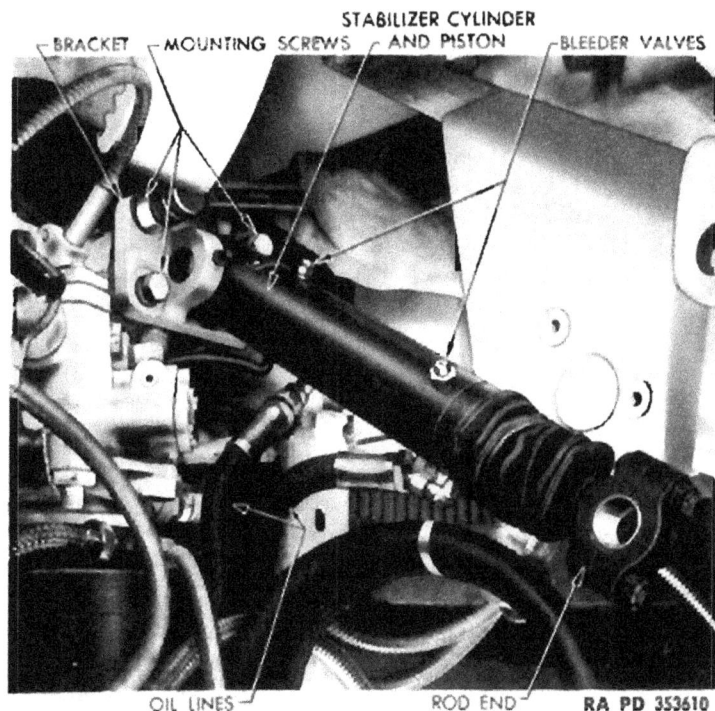

Figure 59. Removing stabilizer cylinder and piston, and bracket (M24).

t. REMOVE DOME LIGHTS AND FIRING SWITCH CABLE HARNESS WITH TELEPHONE BOX. Remove all screws and washers holding clips of wiring harness to tapping blocks on turret roof. Remove wiring harness and box as an assembly.

u. REMOVE COMMANDER'S VISION CUPOLA. Remove 11 screws and washers holding cupola to turret roof. Attach sling to cupola and remove from turret.

Section IV. OVERHAUL OF SUBASSEMBLIES

39. Commander's Vision Cupola

a. DISASSEMBLY. (1) *Remove cupola door assembly.* Open cupola door to the vertical position and hold in this position. Remove eight screws holding cap assemblies to outer ends of hinge (fig. 64). Remove caps. Torsion springs are now released with the door in the vertical position. Pull all six springs out of tube. While supporting door, push tube out of hinge bushing. Lift out door.

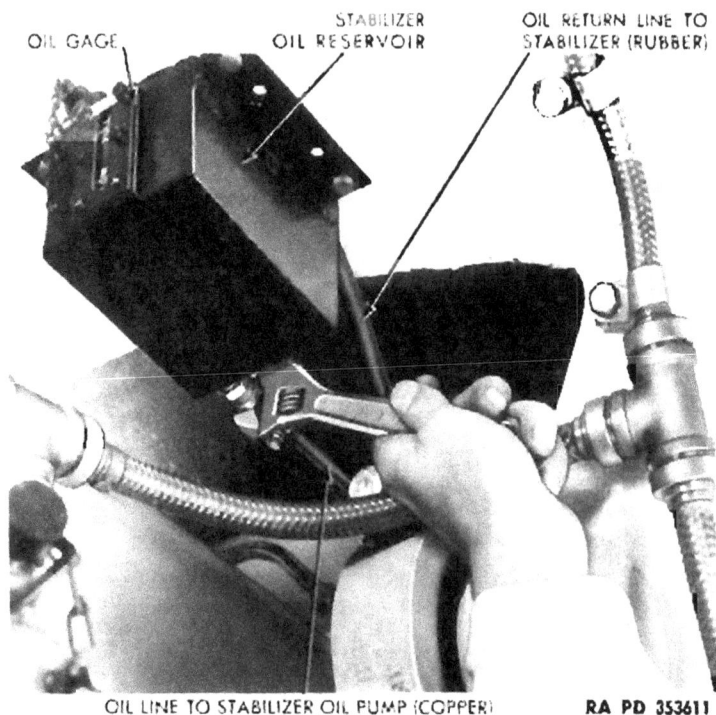

Figure 60. Removing oil line from stabilizer oil pump to oil reservoir (M24).

(2) *Remove door latch.* Remove two screws which fasten cupola door latch spring and handle to bottom of cupola door. Slide out latch clevis plug and remove spring and handle from door.

(3) *Remove vision block.* Remove the two wedge lock screws and the center jack screw locking each block in position (fig. 64). Remove wedge and allow block to drop down. Remove 14 screws holding bezel assembly to cupola. Remove bezel assembly.

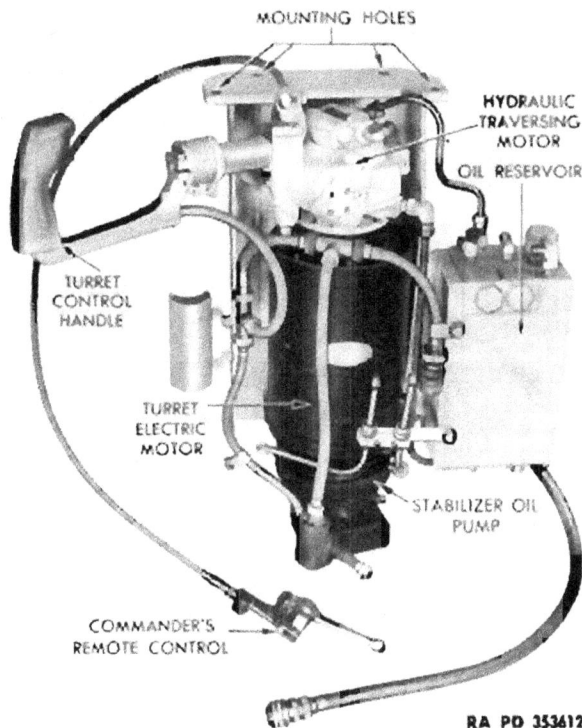

MOUNTING HOLES

HYDRAULIC TRAVERSING MOTOR

OIL RESERVOIR

TURRET CONTROL HANDLE

TURRET ELECTRIC MOTOR

STABILIZER OIL PUMP

COMMANDER'S REMOTE CONTROL

RA PD 353612

Figure 61. Turret motor, oil pumps, and traversing mechanism oil reservoir assembly (M24).

(4) *Remove recess filler block.* Slide recess filler block locking levers toward outside of cupola door and allow block to slide out.

(5) *Remove door race plate.* Remove 15 screws holding door race cover and seal assembly to door race plate. Remove cover. Remove two screws holding bearing race cap to door ring. Remove cap. Place block of wood under door race plate to hold plate in line with door ring. Remove 122 ball bearings from race through the opening at the rear where the bearing race cap was removed.

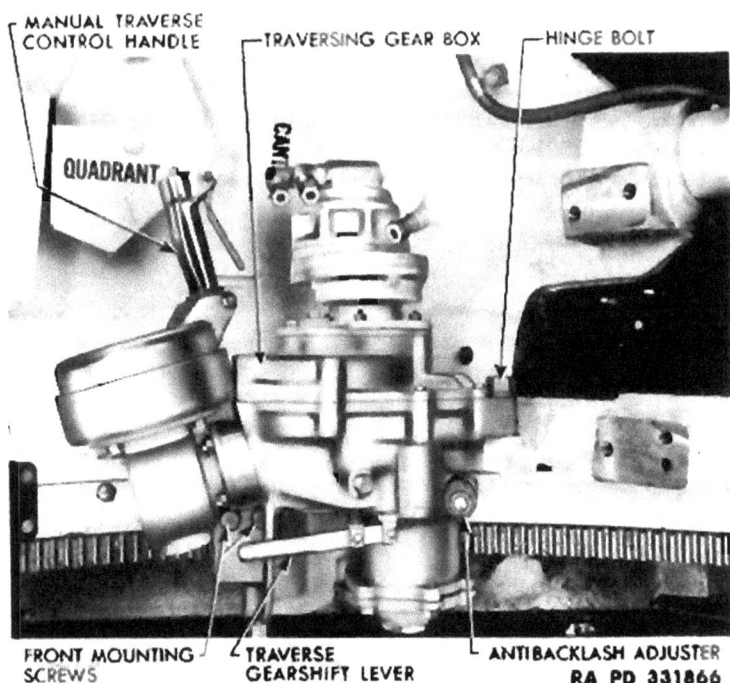

MANUAL TRAVERSE
CONTROL HANDLE — TRAVERSING GEAR BOX — HINGE BOLT

QUADRANT

FRONT MOUNTING — TRAVERSE — ANTIBACKLASH ADJUSTER
SCREWS GEARSHIFT LEVER RA PD 331866

Figure 62. Traversing gear box mounting screws (M24).

Note. Two sizes of balls are used. After ball bearings have been removed, lift door race plate out of door ring.

b. CLEANING. Wash all parts thoroughly in dry cleaning solvent, and dry with compressed air.

c. INSPECTION. (1) *Cupola race balls.* Examine each of the cupola race balls for scratches, chipped or flat spots, cracks, or any other condition which would render them unserviceable. Measure diameter of each ball with micrometer. Diameter of the large ball should be 0.4998 to 0.5002 inch. Diameter of small ball should be 0.4373 to 0.4377 inch. When selecting a single new ball, or a series of new balls to be used with old balls, exercise care to select balls toward low limit, as all old balls will be worn, and a more uniform pressure will result.

(2) *Cupola race rings.* Thoroughly examine ball races in race rings for cracks, scores, and brinell marks caused by balls. Any race ring which has been in service for some time may have brinell marks, but if marks do not exceed 0.005 inch in depth, the ring is satisfactory for further use. If brinell marks exceed 0.005 inch in depth or ring is otherwise damaged, replace or rebuild the complete cupola assembly.

(3) *Vision blocks.* Inspect all vision blocks for discoloration, cracks, or separation. Replace vision blocks which show damage.

(4) *Torsion springs.* Inspect the four torsion springs for cracks or breaks. Replace springs which show damage.

d. ASSEMBLY. (1) *Install door race plate.* Place the door ring on a flat surface and block up the door plate until the ball races in the door and plates line up. Install the 122 ball bearings in the race through the opening at the rear of the race ring.

Note. Install the large and small balls alternately.

After all of the 122 balls have been installed, install the bearing race cap in the door ring and install the 2 attaching screws and lock washers. Place the door race cover and seal assembly over the door and install the 15 attaching screws.

(2) *Install recess filler block.* Hold the two filler block locking levers in the released position and install the recess filler block. Slide the locking levers toward the center of the door to hold the filler block in position.

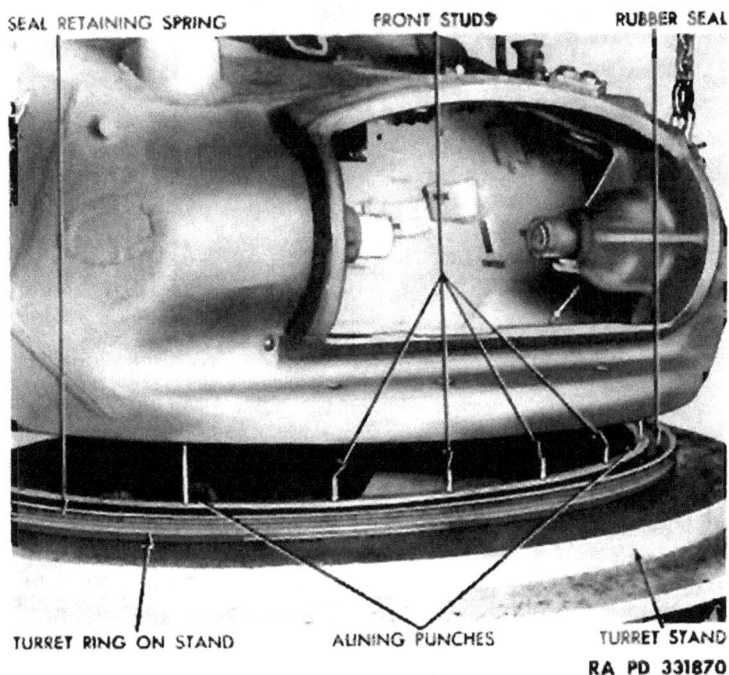

SEAL RETAINING SPRING   FRONT STUDS   RUBBER SEAL

TURRET RING ON STAND   ALINING PUNCHES   TURRET STAND

RA PD 331870

Figure 63. Removing turret from ring gear (M24).

(3) *Install vision blocks.* Place the bezel assembly over the vision openings and install the 14 screws holding bezel to cupola. Slide block into opening from the bottom. Place wedge into position and install jack screw and two wedge lock screws. Repeat this operation for the other five vision blocks.

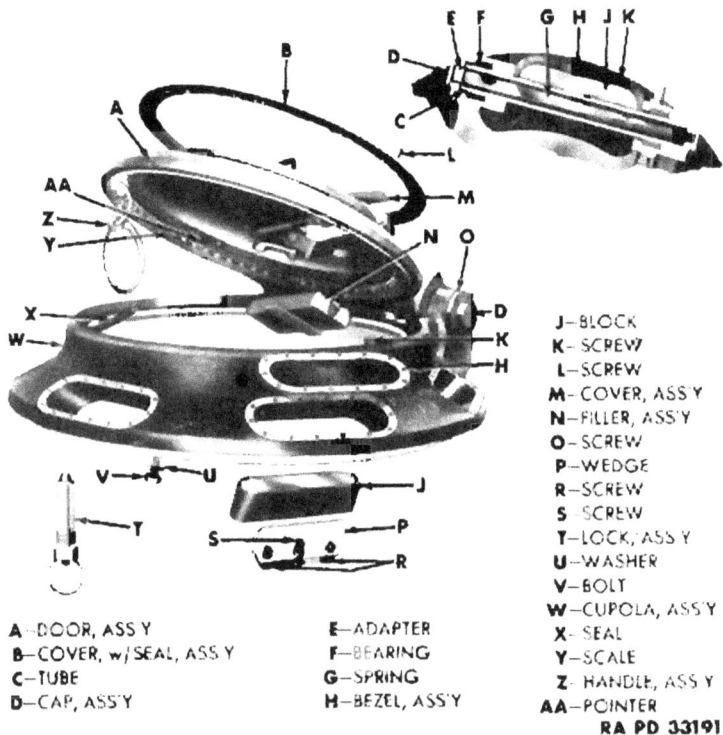

J—BLOCK
K—SCREW
L—SCREW
M—COVER, ASS'Y
N—FILLER, ASS'Y
O—SCREW
P—WEDGE
R—SCREW
S—SCREW
T—LOCK, ASS'Y
U—WASHER
V—BOLT
W—CUPOLA, ASS'Y
X—SEAL
Y—SCALE
Z—HANDLE, ASS'Y
AA—POINTER

A—DOOR, ASS'Y
B—COVER, w/SEAL, ASS'Y
C—TUBE
D—CAP, ASS'Y
E—ADAPTER
F—BEARING
G—SPRING
H—BEZEL, ASS'Y

RA PD 331911

Figure 64. Commander's vision cupola and door (M24).

(4) *Install door latch.* Place spring and latch handle in bracket on door. Place latch clevis plug in bracket and install the two retaining screws.

(5) *Install cupola door assembly.* Place door in vertical position next to hinges on cupola. While holding door vertical, install tube through bushings in hinges. Next, place six torsion springs in tube. Position cap assembly over ends of hinges, making sure that torsion springs fit into cut-out in caps. Install eight screws holding caps to hinges. Close door.

40. Turret Traversing Ball Ring

a. DISASSEMBLY. (1) *Mark outer upper and outer lower ring.* Place turret traversing ball ring on large surface plate. Mark outer upper and outer lower ring with prick punch (fig. 65) to permit locating parts correctly at time of reassembly.

(2) *Remove outer upper ring.* Remove 14 socket head cap screws holding outer upper ring to outer lower ring. Install three ½–20 NF threaded eyebolts in outer upper ring, attach hoist to eye bolts and remove ring (fig. 66).

(3) *Remove ball bearings.* Pry up inner turret ring from outer lower ring and remove the 132 ball bearings and 12 bearing retainers. Lift inner turret ball bearing ring off outer lower ring.

b. CLEANING. Wash all parts thoroughly in dry cleaning solvent and dry with compressed air.

c. INSPECTION. (1) *Inspect outer lower, outer upper, and inner ring bearing races.* Inspect the turret ring gear teeth for damage. Small nicks and burs on gear teeth can be removed by using a fine file. Inspect ball races on rings for brinell marks. Using a dial indicator attached to a plate (fig. 67), check depth of brinell marks. Any race ring which has been in service for some time may have brinell marks,

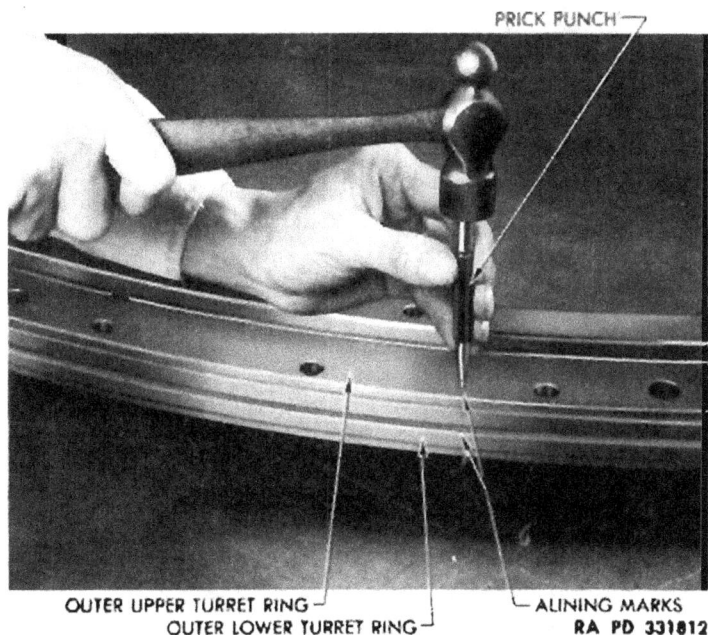

PRICK PUNCH

OUTER UPPER TURRET RING
OUTER LOWER TURRET RING
ALINING MARKS
RA PD 331812

Figure 65. Marking outer upper and outer lower turret ring (M24).

85

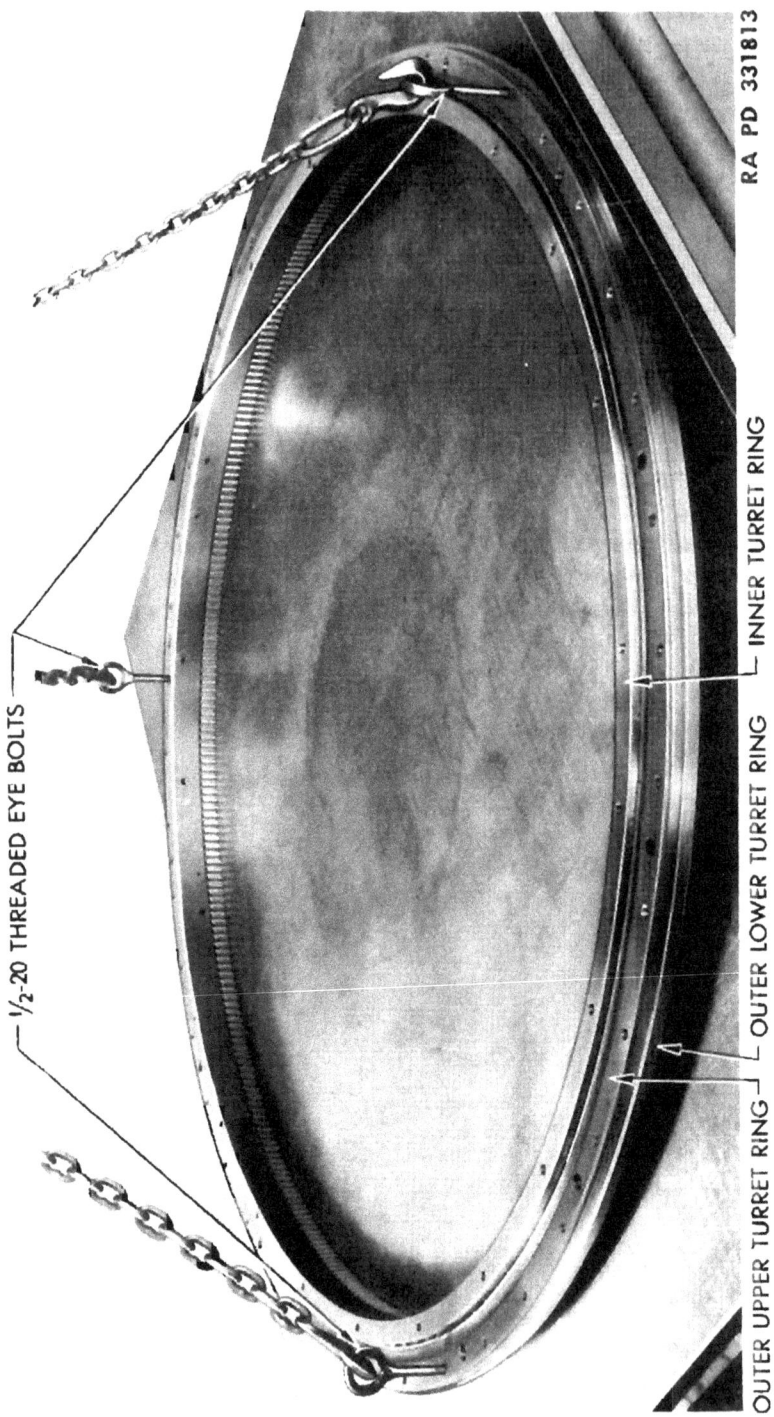

OUTER UPPER TURRET RING — OUTER LOWER TURRET RING — INNER TURRET RING

½-20 THREADED EYE BOLTS

RA PD 331813

Figure 66. Removing outer upper turret ring from outer lower turret ring (M24).

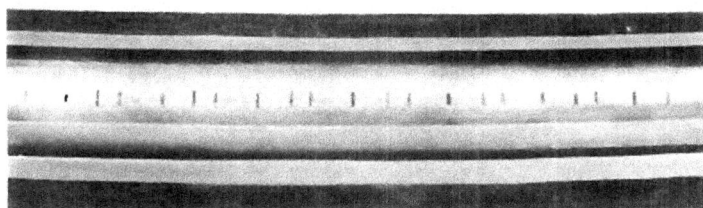

BRINELL MARKS IN UPPER RACE RING

CHECKING LOWER RACE RING   CHECKING UPPER RACE RING

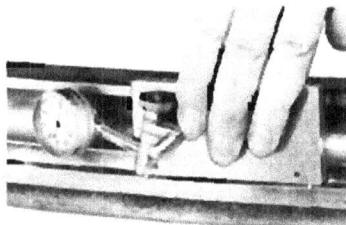

FIXTURE FOR LOWER RACE   FIXTURE FOR UPPER RACE

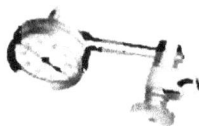

CHECKING FIXTURES AND DIAL INDICATOR

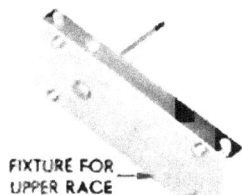

RA PD 309639

Figure 67. Checking depth of brinell marks (M24).

but if marks do not exceed 0.005 inch in depth, the ring is satisfactory for further use. If brinell marks exceed 0.005 inch in depth, or ring is otherwise damaged, replace or rebuild the complete turret ring assembly.

(2) *Inspect turret ball bearings.* Examine each of the turret balls for scratches, chips or flat spots, cracks or any other conditions which would render them unserviceable. Measure diameter of each ball with micrometer. Diameter of each ball should be 0.9995 to 1.0005 inches. When selecting a single new ball, or a series of new balls to be used with the old balls, exercise care to select balls toward the low limit, as all old balls will be worn, and more uniform pressure results.

d. ASSEMBLY. (1) *Install inner ball bearing ring over outer lower ball bearing ring.* Place outer lower turret ball bearing ring and ring gear on large surface plate. Carefully lower the inner turret ball bearing ring into the outer lower ring.

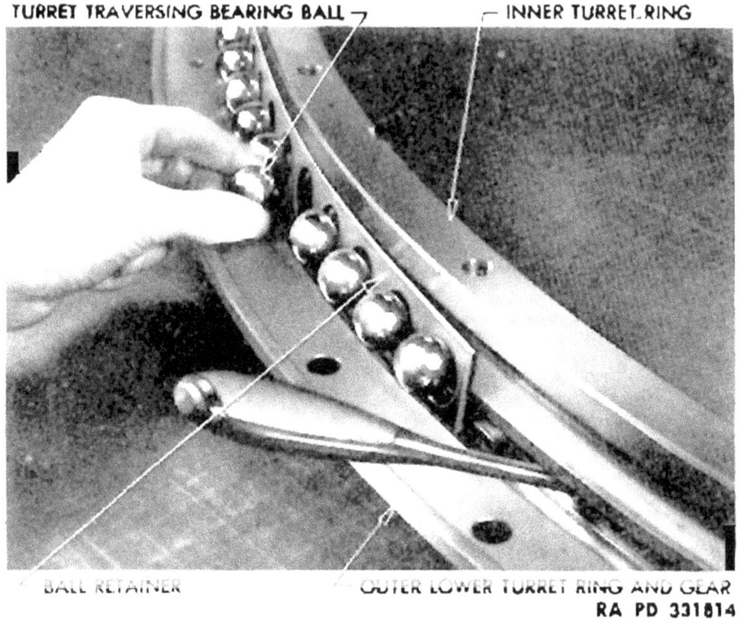
TURRET TRAVERSING BEARING BALL
INNER TURRET RING
BALL RETAINER
OUTER LOWER TURRET RING AND GEAR
RA PD 331814

Figure 68. Installing turret balls (M24).

SOCKET HEAD CAP SCREWS
⅜ HEX WRENCH
OUTER UPPER TURRET RING
OUTER LOWER TURRET RING
C- CLAMP
RA PD 331815

Figure 69. Installing outer upper ring retaining screws (M24).

(2) *Install ball bearings and retainers.* Carefully insert a screwdriver between the inner ball bearing ring and the outer lower ring. With the ring raised in this position, install the turret balls in the retainers (fig. 68).

Note. Coat balls and races with general purpose grease before installation.

After two retainers have been installed, the screwdriver may be removed from this position. Place the other retainers loaded with balls next to the opening and carefully raise up the inner ring until the balls drop into the proper position in the races. Install all balls and retainers in this manner.

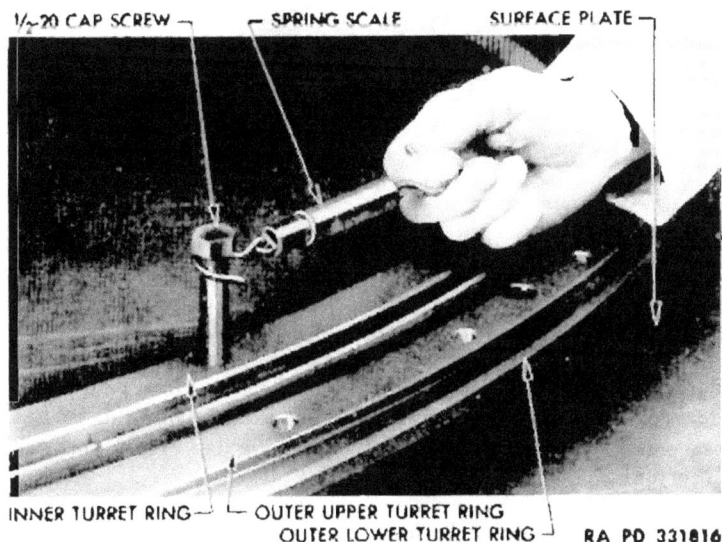

½-20 CAP SCREW —   — SPRING SCALE   SURFACE PLATE —

INNER TURRET RING —   — OUTER UPPER TURRET RING   OUTER LOWER TURRET RING —   RA PD 331816

Figure 70. Checking tightness of bearings (M24).

(3) *Install outer upper turret ball bearing ring.* Install three ½–20 NF eyebolts in the outer upper ring and attach hoist to eyebolts. Carefully lower upper ring over lower ring gear, being sure to line up the punch marks which were made before disassembly (fig. 65). Remove hoist and eyebolts.

(4) *Install lock screws.* Install 14 socket head cap screws and washers holding outer upper ring to lower ring.

Note. Turn all screws just enough to get them started.

Before tightening screws, install 4 C-clamps over the upper and lower rings and clamp rings securely (fig. 69) ; then tighten all screws evenly to a torque tightness of 25–30 foot-pounds.

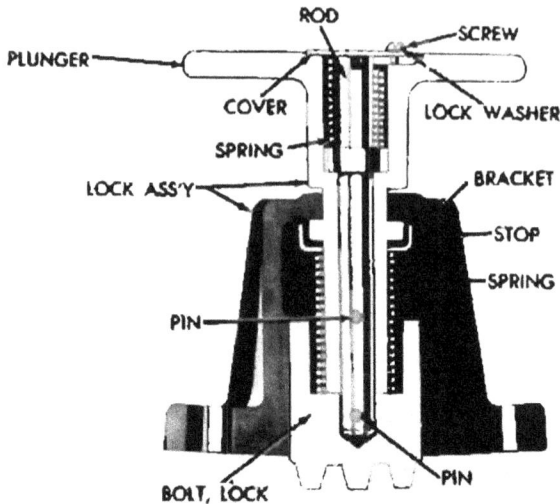

Figure 71. Turret traversing lock—cross-sectional view (M24).

(5) *Check tightness of bearings.* Install a ½–20 NF cap screw in the inner turret ball bearing ring. Using a spring scale, rotate inner ring 360° (fig. 70). Effort required to rotate inner ring must not exceed 12.5 pounds maximum. If effort required to rotate inner ring exceeds 12.5 pounds, loosen screws and retighten evenly until within this limit.

41. Turret Traversing Lock

a. DISASSEMBLY. (1) *Remove lock plunger torsion spring.* Remove two screws holding plunger rod cover to plunger. Remove spring by prying out of plunger (fig. 71).

(2) *Remove lock plunger.* Pull plunger out of bracket, remove lubricating fitting, compressing lock plunger spring (approximately one-half inch) until lock bolt pin can be seen through the lubrication fitting hole, and drive out lock pin. Hold lock bolt in compressed position while removing pin to prevent lock plunger spring from forcing lock bolt out of bracket and losing parts. When pin has been removed, spring will force lock bolt out of bracket. Remove plunger spring, stop, and plunger from bracket (fig. 71).

(3) *Disassemble lock plunger.* Turn plunger rod until lock plunger straight pin can be seen through hole in plunger. Drive out pin. Slide plunger rod out of plunger.

b. CLEANING. Wash all parts thoroughly in dry cleaning solvent and dry with compressed air.

c. INSPECTION. (1) *Inspect bracket.* Inspect bracket for cracks or breaks that may cause failure in service.

(2) *Inspect lock bolt.* Inspect teeth on lock bolt for nicks and burs. Small nicks and burs can be removed by using a fine file. Insert lock bolt in bracket and see that it slides freely. Remove nicks and burs from side of bolt and inside of bracket with fine file.

(3) *Inspect spring.* Check free length of turret lock plunger torsion spring. Free length should be 0.960 inch. Check free length of lock plunger spring. Free length should be 2.888 inches. Compress spring to 1.625 inches and measure pressure while compressed. Pressure should be 19–21 pounds.

d. ASSEMBLY. (1) *Install plunger rod into plunger.* Lubricate lock rod and install in plunger (fig. 71). Locate hole in rod and drive straight pin in flush with shoulder of plunger.

(2) *Install plunger in bracket.* Position plunger spring stop over plunger with concave side toward handle. Place spring over plunger and position lock bolt over end of spring. Compress spring until hole in lock bolt lines up with hole in plunger rod. Install lock bolt straight pin through the lock bolt and rod until pin is flush with lock bolt (fig. 71).

(3) *Install lock plunger torsion spring and cover.* Place bracket assembly in vise with plunger straight up. Insert plunger torsion spring into plunger rod. Turn spring clockwise until top end of spring fits in cut-out in end of plunger. Place plunger rod cover over spring and install two retaining screws and washers.

42. Periscope Linkage

a. DISASSEMBLY. (1) *Remove connector from periscope holder.* Using a small pin punch, drive dowel pin which holds stud in periscope holder bracket, out of bracket. Place a brass drift on end of stud and drive connector and stud assembly out of periscope holder bracket (fig. 72).

(2) *Disassemble connector.* Place end of stud in vise and bend down tang on lock washer. Using a spanner wrench, remove nut and washer from stud. Using a small screw driver, remove snap ring holding the bearing in connector. Press bearing and stud out of connector. Press bearing off of stud.

(3) *Remove link assembly from elevating bracket.* Bend down tang on lock washer, and remove nut and washer from stud. Using a small screw driver, remove snap ring holding the bearing in lower connector of link assembly. Tap link and bearing assembly off stud on elevating bracket. Remove bearing from lower connector.

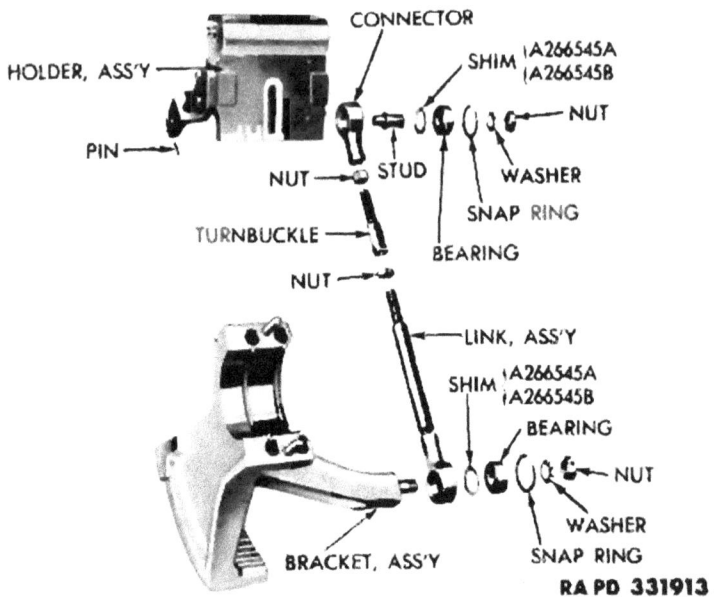

Figure 72. Periscope linkage, disassembled (M24).

b. CLEANING. Wash all parts, with the exception of the sealed bearing, in dry cleaning solvent, and dry with compressed air.

Caution: Do not wash the sealed bearing in dry cleaning solvent.

c. INSPECTION. (1) *Inspect bearings.* Inspect sealed bearings for roughness or other conditions that may cause failure in service. Replace any bearings which show damage. Measure outside diameter of bearing. Outside diameter of bearing should be 1.8499 to 1.8504 inches. Measure inside diameter of bearing. Inside diameter should be 0.7870 to 0.7874 inch.

(2) *Inspect connector.* Inspect connector for cracks or damage that may cause failure in service. Measure diameter of bearing bore in connector. Diameter of bearing bore should be 1.8502 to 1.8508 inches. Inspect thread on inside of connector for damage. Damaged thread can be cleaned by using a ⅝–18NF–2 tap.

(3) *Inspect gun sighting periscope link upper stud.* Inspect gun sighting periscope link upper stud for damage or stripped threads. Measure bearing seat on stud. Diameter of stud should be 0.7875 to 0.7880 inch.

(4) *Inspect link assembly.* Inspect link assembly for being bent. Measure diameter of lower bearing connector. Bearing surface of connector should be 1.8502 to 1.8508 inches. Inspect threads on upper end of link assembly for damage or stripped threads.

(5) *Inspect gun sighting periscope linkage turnbuckle.* Inspect inner and outer threads of turnbuckle for damage. Outer turnbuckle threads can be cleaned by using a ⅜-18NF-3 right-hand thread die.

d. ASSEMBLY. (1) *Assemble link assembly to elevating bracket assembly.* Place lower connector of link assembly over stud on elevating bracket assembly and install shim over stud. Press bearing over stud until the locating groove for snap ring is visible. Install snap ring. Place lock washer over stud and install nut. Tighten nut securely and bend tang of lock washer in slot in nut.

(2) *Assemble top connector to periscope bracket.* Press bearing on stud until bearing contacts shoulder of stud. Place shim in connector and press bearing and stud assembly into connector until the groove for snap ring is visible. Install snap ring. Place lock washer over stud and install nut. Tighten nut securely and bend tang of washer into slot in nut.

(3) *Install connector assembly to periscope holder bracket.* Press stud of connector assembly into hole in bracket, making sure that pin hole in stud lines up with pin hole in bracket. Install a new pin through lock stud and bracket, and peen over ends of pin. The link cannot be assembled to upper connector on periscope holder at this time, as these parts are assembled and locked together at time of bore sighting the 75-mm gun.

Section V. ASSEMBLY OF SUBASSEMBLIES

43. Assembly Procedure

a. INSTALL COMMANDER'S VISION CUPOLA. Coat mounting surface of commander's vision cupola with sealing compound, and lower cupola into position on turret roof. Install 11 screws and washers holding cupola to turret. Wipe off excessive sealing compound.

b. ASSEMBLE DOME LIGHT AND FIRING SWITCH CABLE HARNESS WITH TELEPHONE BOX TO TURRET. Position cable harness over tapping blocks on turret roof, and install screws through clips holding harness assembly to tapping blocks.

c. ASSEMBLE ANTIAIRCRAFT TRIPOD TO TURRET. Position antiaircraft tripod over mounting holes on top of turret, and install six bolts and nuts holding tripod to turret.

d. ASSEMBLE TURRET RING GEAR TO TURRET. Place turret ring gear on mounting stand. Coat outer edge of ring gear with nonvulcanizing rubber cement. Coat inside edge of turret rubber seal with nonvulcanizing rubber cement. Assemble rubber seal over outer edge of ring gear so that feather edge of seal is toward top. Install rubber seal retaining spring around outside edge of rubber seal. Coat feather edge of rubber seal with hydraulic brake fluid. The brake fluid will prevent the rubber seal from sticking to the bottom of the turret and prevent damage to the seal when the turret is rotated. Install four ½–20 NF studs in the front top holes of the ring gear (fig. 62). Attach a sling to lifting eyes on turret and lower turret over ring gear, making sure the 4 studs enter the 4 holes at the front of the turret. Install 4 nuts and washers holding turret to ring gear. Install 16 cap screws (½–20 NF x 1¼) with internal toothed lock washers holding turret to ring gear.

Caution: Be sure screws are not longer than 1¼ inches. Longer screws will not hold turret properly.

e. INSTALL TRAVERSING GEAR BOX IN TURRET. Position traversing gear box over mounting holes in turret. Install two screws in front of manual control operating handle, holding gear box hinge to turret. Install antibacklash stud through mounting hole bracket and into turret ring. Remove two nuts used for installing stud. Position antibacklash spring over stud and install spacer, flat washer, nut, and lock nut. Rotate turret 360° to locate high spot between gear and turret ring. When high spot is located, tighten nut on spring until slight drag is noticeable. Back off nut one fourth turn and lock with lock nut.

f. ASSEMBLE TURRET MOTOR, OIL PUMPS, AND TRAVERSING MECHANISM OIL RESERVOIR ASSEMBLY TO TURRET. Position turret motor, oil pumps, and traversing mechanism oil reservoir assembly up under base of turret and install four screws through turret motor, mounting bracket into turret. Install and connect three tubes from oil gear motor to connections on top of reservoir. Install and connect two oil tubes from top of oil pump to connection on top of reservoir. Install and connect one oil tube from front of oil pump to connection on bottom of reservoir. Install and connect copper tube from stabilizer oil pump to stabilizer oil reservoir on turret roof.

g. ASSEMBLE STABILIZER CYLINDER AND PISTON, AND BRACKET TO TURRET. Position stabilizer cylinder and piston, and bracket over mounting holes in turret, and install housing. Install six screws and nuts holding periscope housing assembly to roof of turret.

h. ASSEMBLE GUNNER'S PERISCOPE HOUSING ASSEMBLY TO ROOF OF TURRET. Place gasket over opening for periscope housing and install housing. Install six screws and nuts holding periscope housing assembly to roof of turret.

i. ASSEMBLE COMMANDER'S AND GUNNER'S BACK REST PADS TO TURRET. Position commander's and gunner's back rest pads over locating holes on turret and install the attaching screws and washers.

j. ASSEMBLE 75-MM. GUN AND GUN MOUNT TO TURRET. Attach a hoist to lifting eyes of gun shield, and raise gun into position in front of opening of turret (fig. 73). Elevate muzzle and carefully maneuver gun through turret opening until bearings on trunnions are properly located. Install bearing caps over trunnion bearings, and install nuts. Tighten nuts to 300 foot-pounds. Raise and lower gun tube several times to see that it operates freely without any bind.

k. INSTALL GUIDE AND EJECTOR MECHANISM. Position guide and ejector mechanism on bottom of gun cradle. Install attaching screws and lock washers.

l. INSTALL FIRING MECHANISM ASSEMBLY. Position 75-mm. firing mechanism assembly over mounting screw holes and install two screws at rear and three screws at right side holding mechanism to recoil cylinder. Connect solenoid and wiring harness to firing mechanism. Position machine gun bracket on right side of 75-mm. gun, and install four cap screws attaching front of bracket, to trunnion bearing cap, and install four cap screws attaching rear of machine gun bracket to side of 75-mm. gun. Reconnect wires to machine gun firing solenoid.

m. INSTALL BREECH GUARD AND TELESCOPE MOUNTING BRACKET. Place breech guard in the mounting holes in the recoil cylinder and install large nuts and washers holding guard to cylinder. Position telescope mounting bracket over mounting holes and install attaching screws and washers.

n. ASSEMBLE COMMANDER'S AND GUNNER'S SEAT TO TURRET. Place commander's and gunner's seat assembly over mounting holes on turret and turret support motor bracket, and install attaching screws and lock washers.

o. ASSEMBLE STABILIZER MOUNTING BRACKET TO RECOIL CYLINDER. Position the stabilizer mounting bracket to the bottom of the recoil cylinder, and install four attaching screws and washers.

p. INSTALL STABILIZER. Position stabilizer against stabilizer mounting bracket and install four attaching screws and lock washers. Connect electrical conduit to the stabilizer by inserting the multiprong plug into the back of the stabilizer.

q. INSTALL TURRET TRAVERSING AND TRAVEL LOCKS. Position turret traversing lock on front of turret ball ring, and install two attaching screws and washers. Check lock to see that it operates freely in both positions. Position the gun travel lock on top of recoil cylinder and install connecting pin and cotter pin.

r. BALANCING THE GUN AND MOUNT. (1) Install all parts not previously mentioned. This includes the telescope and light; elevation quadrant and light caliber .30 empty cartridge bag; 250-round box of caliber .30 ammunition, or its equivalent, in the ammunition box holder under the machine gun. Load a 75-mm. round (inert) in the gun. Place the breech operating lever in its bracket on the rear of the recoil guard. Disconnect the stabilizer piston from the mount, and hang a 5¾-pound weight on the stabilizer piston rod pin. Disengage the elevating mechanism.

(2) Drop small counterweights B7051216 in the space provided at the rear end of the recoil guard. If seven of these counterweights are not sufficient to balance the gun, hook the large counterweight D7051429 on the rear end of the guard and remove enough small counterweights as necessary to balance the gun. Place the gun at 0° elevation. Hook a spring scale in the muzzle of the gun. The effort required to elevate the gun should be within 1 pound of the effort required to depress the gun. Bolt the counterweights in position and recheck the balance.

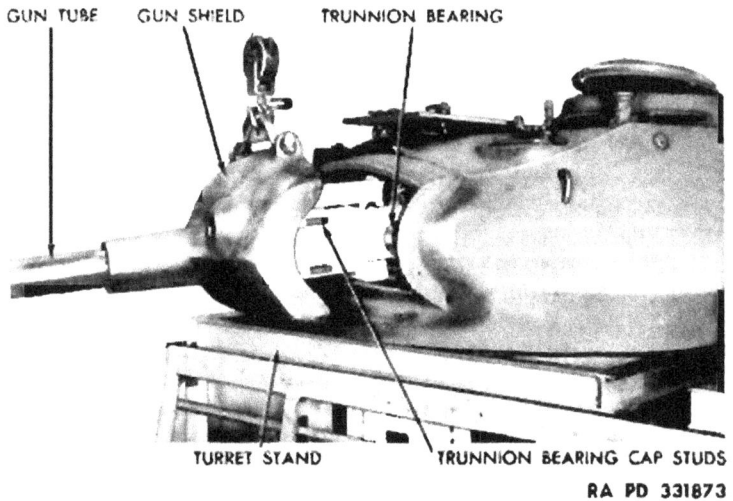

Figure 73. Installing 75-mm. gun and shield (M24).

a. INSTALL ELEVATING MECHANISM. If the elevating mechanism has been used previously on the same gun mount, assemble the pinion in its original position relative to the sector teeth. Reinstall the original shims, assemble the elevating mechanism bracket over the original dowels, and tighten the attaching screws. Check the clearance and the tooth contact between the elevating pinion and sector by passing 0.0015-inch brass shim stock, 1¾ to 2 inches wide, between the pinion and sector teeth. This can be performed by operating the elevating mechanism crank. The clearance should be such that at least one thickness of this stock can be passed between the teeth without binding. Not more than two thicknesses are required to secure a good tooth contact impression on the shims. If, however, more than two pieces of brass shim stock are required to secure a definite impression, add shims between the elevating mechanism supporting

96

bracket and the turret to move the elevating mechanism toward the rear, then tighten the elevating mechanism bracket screws. If the tooth impression on the brass shim stock shows contact on the ends of the teeth only, it will be necessary to reposition the elevating mechanism. Drill out the upper elevating mechanism bracket dowel, tighten the attaching screws only enough to hold the mechanism in position, and then rotate the elevating mechanism and bracket slightly by tapping with a hammer. After correct pinion, sector clearance, and good tooth contact have been secured, tighten the bracket attaching screws securely, but do not install the upper dowel until after the friction measurements described below have been completed. If a new gun mount or new elevating mechanism is to be installed, pull elevating mechanism toward rear so as to secure substantially zero backlash between the elevating pinion and sector teeth. Slip shims between the elevating mechanism support bracket and the turret to hold the support bracket toward the rear. Make sure that the same number and thickness of shims are placed under each of the support pads. Adjust the pinion clearance and tooth contact and tighten the bracket screws. Using a spring scale and wire loop at the muzzle, and with the elevating mechanism disengaged, measure the effort required to elevate and depress the gun throughout its elevation range. The force required should not exceed the figures given in the following table:

	Pounds
To start gun from zero elevation (up or down)	4
To move gun to maximum elevation	7
To start gun down from maximum elevation	2
To move gun to maximum depression	6
To start gun up from maximum depression	3

If the friction is excessive, check the following in the order listed:

(1) Make certain that there are no interferences between the gun mount tipping parts and fixed turret parts.

(2) Make certain there are no damaged elevating sector or pinion teeth.

(3) Check periscope linkage for binding.

(4) Check the trunnion bearing dust covers to make certain that they are not rubbing against the bearings.

(5) Increase elevating mechanism clearance by changing shims so as to move elevating mechanism forward 0.002 inch, making certain this does not produce excessive lost motion by checking with brass shim stock as described previously. Backlash should be 0.000 to 0.003 inch for satisfactory operation.

t. INSTALL ELEVATING QUADRANT. Place elevating quadrant and bracket over mounting holes at the rear end of stabilizer cylinder and piston bracket. Install three screws and washers holding quadrant to bracket.

u. BLEED HYDRAULIC SYSTEM IN TURRET. Remove plug from top of traversing mechanism oil reservoir, and fill with hydraulic oil until it is seen through the glass gage. Reinstall the plug. Fill stabilizer oil reservoir until it is at least two-thirds full. Remove screw and connect rubber hose to top bleeder valve on stabilizer cylinder (fig. 74). Place end of hose in clean container. Loosen top bleeder valve and pull stabilizer piston up to top and close top bleeder valve. Disconnect hose from top bleeder valve, remove screws, and connect hose

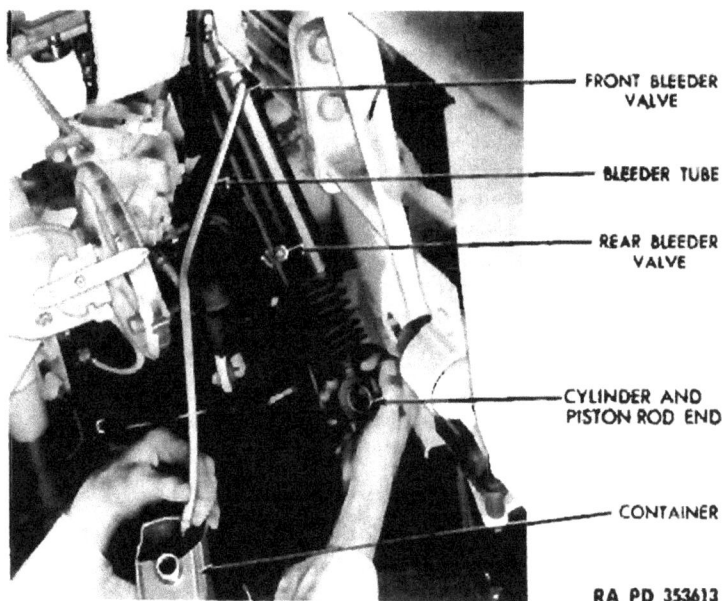

FRONT BLEEDER VALVE

BLEEDER TUBE

REAR BLEEDER VALVE

CYLINDER AND PISTON ROD END

CONTAINER

RA PD 353613

Figure 74. Bleeding stabilizer cylinder (M24).

to cylinder bottom bleeder valve. Open bottom valve and shove piston all the way down. Repeat above operation until oil is free of air. Recheck oil level in stabilizer oil reservoir and see that it is at least two-thirds full. Take care during bleeding operation to prevent any particles of dirt from entering system.

v. ASSEMBLE STABILIZER CYLINDER AND PISTON TO 75-MM. GUN CRADLE. Raise rear end of stabilizer cylinder and piston until bushing in piston lines up with mounting hole in bracket at bottom of gun cradle. Install pin through the bracket and bushing in cylinder. Tighten the two lock screws and lock nuts holding pin in bracket.

w. TEST TURRET. Place turret assembly in test stand, and connect turret control box to turret motor. Test turret traversing mechanism for freeness and proper operation. Check all oil tubes for leaks, and correct if necessary.

x. SYNCHRONIZATION OF PERISCOPE SIGHTING DEVICES. (1) *Equipment required.* (*a*) Gunner's quadrant.

(*b*) L-shaped leveling block for periscope holder.

(*c*) Bore sighting equipment.

(*d*) Testing target for vehicle being checked.

(2) *Synchronizing procedure.* (*a*) Track the gun along the left vertical line throughout the entire elevation of the gun, using the bore sighting equipment for the gun to determine the gun center line. Adjust the deflection knob of the periscope, mounted in the gunner's periscope holder, to coincide with the right vertical line. Track the periscope along the right vertical line throughout the entire elevation of the gun, using the zero cross hair of the telescope, in the periscope, as the periscope center line.

(*b*) Remove the periscope from the periscope holder and insert the L-shaped leveling block. The purpose of the L-shaped leveling block is to hold the gunner's quadrant at right angles to the inner surface of the clamping face of the periscope holder.

(*c*) Set the gunner's quadrant at zero elevation, place on leveling blocks of gun breech ring (or level surface providing same spot is always used) and elevate or depress gun until bubble is level. Adjust bubble of elevation quadrant M9 until it is level. Loosen screws on scale of elevation quadrant, and set to zero, making sure the scale is still zeroed after the screws have been tightened.

(*d*) With the gun position unchanged, place the gunner's quadrant (set at zero) on the L-shaped leveling block in the periscope holder. If the bubble of the gunner's quadrant is not centered, lengthen or shorten the periscope link enough to center the bubble after loosening the two jam nuts. Elevate the gun to a reading of 320 mils, using the elevation quadrant M9.

Note. Elevate the gun on the light tank M24 to its maximum elevation.

The gunner's quadrant on the L-shaped leveling block in the periscope holder then reads the same as the elevation quadrant, plus or minus 1 mil. If the reading on the periscope holder differs more than plus or minus 1 mil, shorten or lengthen the periscope linkage arm to compensate for the difference. Return the gun to zero reading and if the periscope holder depressed exactly the same number of mils, plus or minus 1 mil, tighten the two jam nuts on the periscope linkage arm, and check to assure that the adjustment was not disturbed when the jam nuts were tightened.

(*e*) With the gun at zero elevation, the periscope holder should be perpendicular to the gun, i. e., the gunner's quadrant on the L-shaped leveling block should read zero. If the periscope holder on these vehicles is not within plus or minus 15 mils of being perpendicular to the gun, recheck the periscope holder and linkage arm.

(f) Recheck the adjustment at zero, 160 to 320 mils elevation, to determine that the periscope and gun move an equal amount.

Note. Elevate the gun on the light tank M24 to its maximum elevation. After final check, wire and seal linkage arm with connector.

Section VI. INSTALLATION OF TURRET

44. Installation Procedure

a. INSTALL TURRET ASSEMBLY. Aline center line mark, which is located just to rear of azimuth indicator mounting holes on inner ball bearing ring, with corresponding mark on ring gear tooth. Lock ball bearing ring in this position with turret traversing lock. Attach turret sling (41-S-3832-54) to lifting eyes on turret (fig. 52), and raise turret into position over opening on hull roof. Aline center line of gun to point forward with center line of hull. Insert two long drift punches through the turret ring attaching screw holes in the turret roof. Carefully lower turret assembly to line up screw holes in ring gear and hull roof. Install 42 screws holding turret ball ring to hull roof. Remove sling from turret.

b. CONNECT FEED CABLES. Connect the three feed cables to front of turret control box on hull floor. Tighten knurled nuts securely. Connect radio ground strap to turret control box.

c. INSTALL AZIMUTH INDICATOR. Place azimuth indicator over the attaching screw holes, carefully meshing teeth on indicator with ring gear. Install the two screws holding indicator to turret. Insert shims between indicator bracket and ball bearing to give proper mesh of split pinion in ring gear.

d. TEST TURRET. Start both engines and set speed at about 2,000 revolutions per minute. Turn on turret motor switch and operate turret. Turret should operate freely in all directions. Check for oil leaks and correct if necessary.

CHAPTER 8

HOWITZER MOUNT (M41 ONLY)

Section I. REMOVAL OF HOWITZER AND MOUNT

45. Removal of Howitzer, Cradle, and Top Carriage

a. GENERAL. The howitzer, cradle, and top carriage are removed separately when stripping the mount from the hull. The work is best performed with special tools for maintenance and repair, as listed in TM 9–1331. TM 9–1331 covers material which is essentially the same as this mount though of a different designation. For recoil oil and equipment for establishing or exhausting oil reserve, refer to TM 9–331.

b. DISENGAGE TRAVELING LOCK. Loosen the traveling lock at muzzle end of howitzer tube, elevate the tube slightly, and fold the lock down against the hull.

c. DISCONNECT HOWITZER ELEVATING ELECTRICAL CABLES. Throw safety switch on howitzer elevating control relay box, and master switch in driver's compartment to OFF position, loosen 10 screws, and remove the cover from the relay box located on the left side of the bottom carriage (fig. 75). Disconnect the three cables running from the electric elevating mechanism to the elevating control relay box, taking precaution to properly identify the wires and terminals for reassembly (fig. 76).

d. DISCONNECT LIMIT SWITCH. Disconnect the limit switch from the equilibrator by removing the cotter pin and drawing the pin from the link (fig. 77).

e. REMOVE GUN MOUNTING STRUCTURE REMOVABLE SECTION. Remove the six cap screws and lock washers which secure the removable section (fig. 78) and remove from the vehicle. Disengage the traversing locks.

f. REMOVAL OF HOWITZER TUBE, CRADLE, AND TOP CARRIAGE. Subsequent operations in the removal of howitzer tube, cradle, and top carriage from the bottom carriage are given in detail in TM 9–1331.

46. Installation of Top Carriage, Cradle, and Howitzer

a. GENERAL. Clean the parts of the mount thoroughly before assembly to the hull. Grease fittings should be open and operative. Pack the pintle bearings with general purpose grease. Lay a film of grease around the belleville springs and in the pintle bearing recess in the top and bottom carriage.

Note. Although this manual is not concerned with repair of the mount, it is advisable to examine bronze bushings in recess before assembly.

Belleville springs ring when suspended and struck with a hard object. Cracked springs will not ring. Put a film of grease on the brass

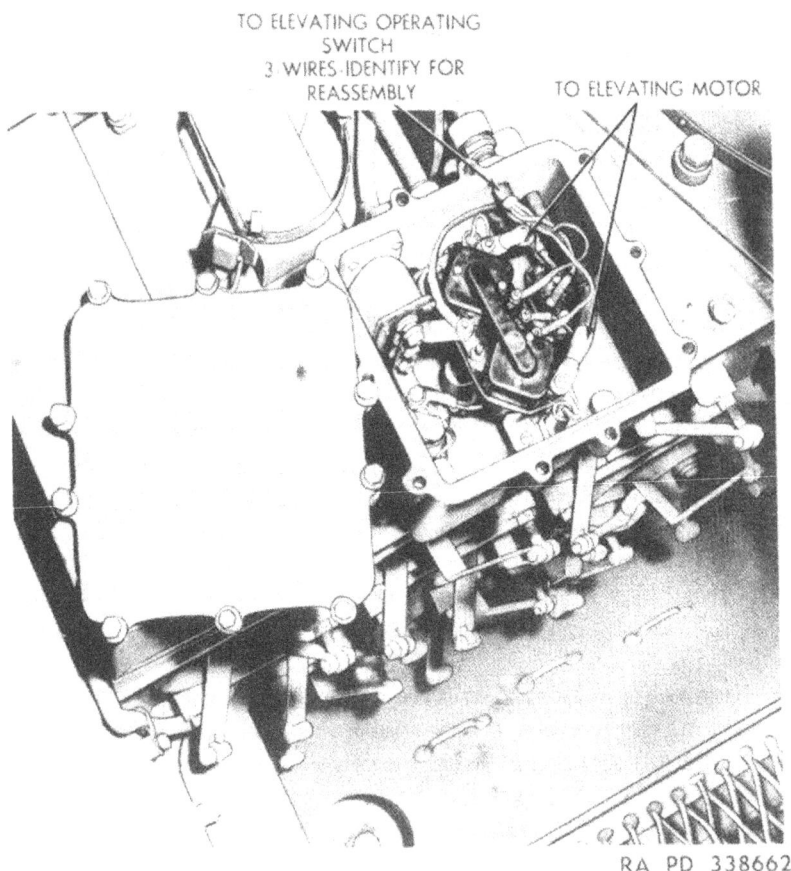

TO ELEVATING OPERATING
SWITCH
3 WIRES-IDENTIFY FOR
REASSEMBLY

TO ELEVATING MOTOR

RA PD 338662

Figure 75. Elevating control relay box with cover removed (M41).

102

liner, on the bottom carriage, on the inside diameter of the upper and lower pintle bushings, and over the felt washer and ring around pintle bearing hole in the top carriage.

b. ASSEMBLE TOP CARRIAGE, CRADLE, AND HOWITZER TO VEHICLE. Refer to TM 9–1331 and TM 9–331.

CABLE 158 TERMINAL F   CABLE 161 TERMINAL Q   CABLE 163 TERMINAL CR

RA PD 338663

Figure 76. Electric control cables disconnected (M41).

c. INSTALL GUN MOUNTING REMOVAL SECTION. Slide the removable section into place under the bottom carriage in rear compartment of vehicle (fig. 78).

d. LIMIT SWITCH. Attach the limit switch to the left equilibrator by installing the pin through the clevis of the connecting link. Secure with a cotter pin (fig. 77).

103

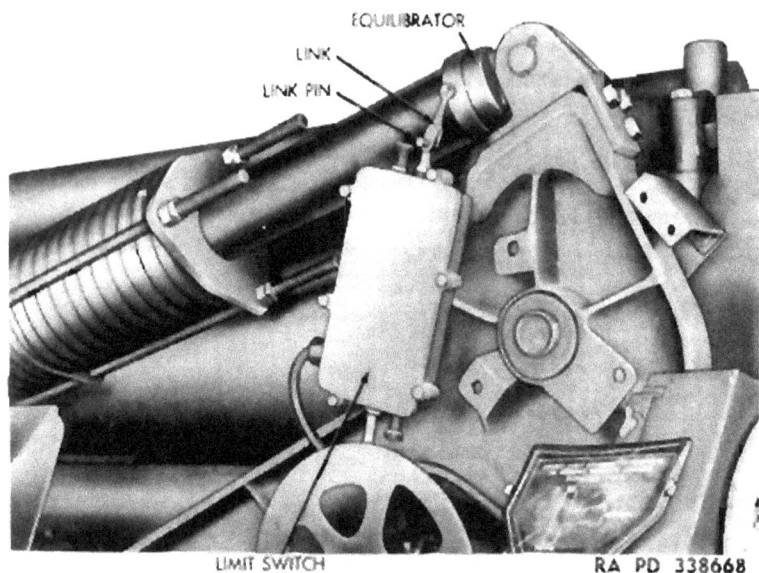

EQUILIBRATOR
LINK
LINK PIN
LIMIT SWITCH
RA PD 338668

Figure 77. Elevating limit switch (M41).

TRAVERSING LOCKS
GUN MOUNTING STRUCTURE
REMOVABLE SECTION
RA PD 338675

Figure 78. Gun mounting structure removable section (M41).

e. ELEVATING CONTROL RELAY BOX. Assemble the cables to their proper terminals in the elevating control relay box, marked for identification at disassembly (fig. 76). Assemble cover and rubber gasket to the box, and tighten the ten cap screws. After connections to the battery have been made, turn on the master switch, located on the box. Check the operations of the electric elevating mechanism by raising and lowering the howitzer with the electric elevating mechanism. Throw switch on control relay box to "OFF" position.

f. LOCKING HOWITZER AT TRAVELING POSITION. Set the howitzer at zero traverse, lower the tube into the traveling lock, and tighten the lock around the tube. Engage the traversing locks at the rear of the mount.

CHAPTER 9
SERVICEABILITY STANDARDS

47. Track Suspension

Point of measurement   *Dimensions of new part*

    a. TRACK PIN.

Diameter	0.827 to 0.828 in.
Straight within	0.004 in.

    b. TRACK SHOE.

Diameter of pin opening	0.829 to 0.835 in.
Clearance between pin and shoe	0.001 to 0.008 in. (loose).
Inside diameter of bushing opening	1.365 to 1.370 in.
Thickness of shoe at grouser	2¼ in.

(Replace if worn beyond 1⅞ inches)

    c. TRACK PIN BUSHING.

Inside diameter	0.830 to 0.832 in.
Clearance between pin and bushing	0.002 to 0.005 in. (loose).

    d. COMPENSATING WHEEL DISK.

Diameter of mounting hole	0.776 to 0.786 in.

    e. COMPENSATING WHEEL HUB AND SPINDLE.

Diameter of outer cup seat	3.6695 to 3.6710 in.
Outside diameter of outer cup	3.6718 to 3.6728 in.
Press fit of cup in seat	0.0008 to 0.0033 in. (tight).
Diameter of inner cup seat	4.4345 to 4.4365 in.
Outside diameter of inner cup	4.4375 to 4.4385 in.
Press fit of cup in seat	0.001 to 0.004 in. (tight).
Diameter of spindle for inner cone	2.4993 to 2.4998 in.
Inside diameter of inner cone	2.5000 to 2.5005 in.
Fit of cone on spindle	0.0002 to 0.0012 in. (loose).
Diameter of spindle for outer cone	1.9993 to 1.9998 in.
Inside diameter of outer cone	2.0000 to 2.0005 in.
Fit of cone on spindle	0.0002 to 0.0012 in. (loose).

    f. COMPENSATING WHEEL ARM AND SPINDLE.

Diameter of spindle for shield	3.365 to 3.367 in.
Inside diameter of shield	3.368 to 3.370 in.

Point of measurement	Dimensions of new part
Fit of shield on spindle	0.001 to 0.005 in. (loose).
Inside diameter of bearing spacer	3.0000 to 3.0005 in.
Diameter of spindle for spacer	3.001 to 3.002 in.
Fit of spacer on spindle	0.0005 to 0.0020 in. (tight).
Diameter of spindle for roller bearing	2.9982 to 2.9992 in.
Inside diameter of roller bearing	2.9994 to 3.0000 in.
Fit of bearing on spindle	0.0002 to 0.0018 in. (loose).
Diameter of spindle for ball bearing	2.3604 to 2.3614 in.
Inside diameter of ball bearing	2.3616 to 2.3622 in.
Fit of bearing on spindle	0.0002 to 0.0018 in. (loose).
Width of ring groove on spindle	0.130 to 0.135 in.
Width of snap ring	0.124 to 0.125 in.
Clearance between ring and groove	0.005 to 0.011 in.
Diameter of roller bearing bore in arm	4.5005 to 4.5020 in.
Outside diameter of roller bearing	4.4992 to 4.5000 in.
Fit of bearing in arm	0.0005 to 0.0028 in. (loose).
Diameter of ball bearing bore in arm	4.3310 to 4.3325 in.
Outside diameter of ball bearing	4.3301 to 4.3307 in.
Fit of bearing in arm	0.0003 to 0.0024 in. (loose).

g. COMPENSATING LINK.

Diameter of bearing bore	3.183 to 3.185 in.
Outside diameter of bearing	3.190 to 3.194 in.
Fit of bearing in bore	0.007 to 0.009 in. (tight).
Diameter of link bolt	1.1225 to 1.1235 in.
Inside diameter of link bearing	1.1245 to 1.1255 in.
Fit of bolt in bearing	0.001 to 0.003 in. (loose).

h. TRACK WHEEL, HUB, AND SUPPORT ARM AND HOUSING.

(1) Track wheel.

Size	25½ x 4½ in.
Tread	Smooth.
Diameter of mounting hole	0.766 to 0.776 in.
Outside diameter of stud	0.742 to 0.750 in.
Fit of wheel on stud	0.016 to 0.034 in. (loose).

(2) Track wheel hub.

Diameter of outer cup seat	3.6695 to 3.6710 in.
Outside diameter of outer cup	3.6718 to 3.6728 in.
Fit of cup in seat	0.0008 to 0.0033 in. (tight).
Diameter of inner cup seat	4.4345 to 4.4365 in.
Outside diameter of inner cup	4.4375 to 4.4385 in.
Fit of cup in seat	0.001 to 0.004 in. (tight).
Diameter of spindle for inner cone	2.4993 to 2.4998 in.
Inside diameter of inner cone	2.5000 to 2.5005 in.
Inside diameter of slinger	5.091 to 5.094 in.

Point of measurement	Dimensions of new part
Diameter of slinger seat on hub	5.095 to 5.098 in.
Fit of slinger on hub	0.001 to 0.007 in. (tight).
Fit of cone on spindle	0.0002 to 0.0012 in. (loose).
Diameter of spindle for outer cone	1.9993 to 1.9998 in.
Inside diameter of outer cone	2.0000 to 2.0005 in.
Fit of cone on spindle	0.0002 to 0.0012 in. (loose).

(3) *Support arm and housing.*

Diameter of oil seal seat in housing	5.0005 to 5.002 in.
Outside diameter of oil seal	5.002 to 5.006 in.
Fit of oil seal in housing	0.0000 to 0.0055 in. (tight).
Diameter of bearing seats in housing	5.0005 to 5.002 in.
Outside diameter of roller bearings	4.999 to 5.000 in.
Fit of bearings in housing	0.0005 to 0.0030 in. (loose).
Outside diameter of support arm shaft	3.4977 to 3.4987 in.
Inside diameter of roller bearings	3.4992 to 3.5000 in.
Fit of bearings on shaft	0.0005 to 0.0023 in. (loose).
Inside diameter of outer bearing spacer	3.500 to 3.502 in.
Fit of spacer on shaft	0.0013 to 0.0043 in. (loose).
Inside diameter of inner bearing spacer	3.505 to 3.510 in.
Fit of spacer on shaft	0.0063 to 0.0123 in. (loose).
Width of snap ring groove in housing	0.155 to 0.160 in.
Thickness of snap ring	0.154 to 0.156 in.
Clearance between ring and groove	0.001 to 0.006 in. (loose).

i. TORSION BAR AND ANCHOR.

(1) *Torsion bar.*

Diameter at arm end	2.442 to 2.444 in.
Diameter at anchor end	2.341 to 2.343 in.

Note. Refer to paragraph 19c for method of measuring diameter of torsion bars and inside diameter of torsion bar anchors.

Diameter at center (right No. 1, 2; left No. 1, 2).	1.719 to 1.734 in.
Diameter at center (right No. 3, 4, and 5; left No. 3, 4, and 5).	1.563 to 1.578 in.
Backlash between mating part	0.004 to 0.008 in.

(2) *Torsion bar spring test.*

Part No.	Position	Angle of twist	Torque (ft-lb) approx.
D60417A	Normal	24°50′	3,200 to 3,600
D60417B	Normal	24°50′	3,200 to 3,600
D60417A	High	60°	8,200 to 8,600
D60417B	High	60°	8,200 to 8,600
D60591A	Normal	15°	2,800 to 3,200
D60591B	Normal	15°	2,800 to 3,200
D60591A	High	50°20′	9,500 to 10,500
D60591B	High	50°20′	9,500 to 10,500

(3) *Torsion bar anchor.*

Inside diameter of torsion bar anchor____ 1.915 to 1.917 in.

Note. Refer to paragraph 12c for method of measuring diameter of torsion bars and inside diameter of torsion bar anchors.

Anchor retaining screw holes_____ $\frac{3}{8}$–24NF–2.

Backlash between mating part_____ 0.004 to 0.008 in.

j. Shock Absorbers. No limits or specifications are provided on shock absorber parts. Each time a shock absorber is disassembled install all of the new parts contained in Kit No. G200–7038157.

k. Track Roller Support Bracket and Spindle.

Diameter of track support bracket mount- $\frac{17}{32}$ in.
     ing holes.

Diameter of spindle at shedder_____ 1.6895 to 1.6900 in.

Diameter of spindle for inner cone_____ 1.6868 to 1.6873 in.

Inside diameter of inner cone_____ 1.6880 to 1.6885 in.

Fit of cone on spindle_____ 0.0007 to 0.0017 in. (loose).

Diameter of spindle for outer cone_____ 1.2493 to 1.2498 in.

Inside diameter of outer cone_____ 1.2500 to 1.2505 in.

Fit of cone on spindle_____ 0.0002 to 0.0012 in. (loose).

Diameter of spindle for adapter_____ 1.6895 to 1.6900 in.

Inside diameter of adapter_____ 1.6873 to 1.6878 in.

Fit of adapter on spindle_____ 0.0017 to 0.0027 in. (tight).

Outside diameter of adapter for spacer__ 2.125 to 2.126 in.

Inside diameter of spacer_____ 2.129 to 2.139 in.

Fit of spacer on adapter_____ 0.003 to 0.014 in. (loose).

l. Support Roller Hub.

Diameter of inner cup seat in hub_____ 3.1225 to 3.124 in.

Outside diameter of inner cup_____ 3.125 to 3.126 in.

Fit of cup in seat_____ 0.0010 to 0.0035 in. (tight).

Diameter of outer cup seat in hub_____ 2.7155 to 2.7165 in.

Outside diameter of outer cup_____ 2.717 to 2.718 in.

Fit of cup in seat_____ 0.0005 to 0.0025 in. (tight).

Diameter of retainer seat in hub_____ 3.187 to 3.189 in.

Outside diameter of retainer_____ 3.191 to 3.194 in.

Fit of retainer in hub_____ 0.002 to 0.007 in. (tight).

m. Track Support Roller Disk.

Diameter of mounting hole_____ 0.468 to 0.484 in.

Size_____ 11 x 3.

Tread_____ Smooth.

Run-out not to exceed_____ $\frac{1}{16}$ in. (total reading).

48. Hull

Point of measurement   *Dimensions of new part*

    a. DRIVER'S DOOR.

Diameter of hinge bushing opening	1.840 to 1.842 in.
Outside diameter of hinge bushing	1.844 to 1.845 in.
Fit of bushing in door	0.002 to 0.005 in. (tight).
Inside diameter of hinge bushing	1.250 to 1.253 in.
Diameter of hinge pin	1.246 to 1.248 in.
Clearance between pin and bushing	0.002 to 0.007 in.

    b. DOOR HINGE YOKE.

Diameter of hinge pin opening (keyway end).	1.248 to 1.249 in.
Diameter of hinge pin opening (plain end).	1.249 to 1.250 in.
Diameter of hinge pin	1.246 to 1.248 in.
Fit of pin in yoke	0.000 to 0.003 in. (tight).
Diameter of bearing seat on yoke	3.1494 to 3.1504 in.
Inside diameter of bearing	3.1490 to 3.1496 in.
Fit of bearing on yoke	0.0002 to 0.0014 in. (tight).
Width of ring groove in yoke	0.162 to 0.0172 in.
Width of external ring	0.154 to 0.158 in.
Clearance between ring and groove	0.004 to 0.018 in.

    c. DOOR HINGE BASE.

Diameter of bearing bore in base	4.9211 to 4.9223 in.
Outside diameter of bearing	4.9205 to 4.9213 in.
Clearance between bearing and base	0.0002 to 0.0018 in.
Width of ring groove in base	0.162 to 0.172 in.
Width of internal ring	0.154 to 0.158 in.
Clearance of ring in groove	0.004 to 0.018 in.

49. Turret (M24)

    a. COMMANDER'S VISION CUPOLA.

      (1) *Cupola race balls.*

Diameter of large ball	0.4998 to 0.5002 in.
Diameter of small ball	0.4373 to 0.4377 in.
Number of large balls	60 or 61. (See note.)
Number of small balls	60 or 61. (See note.)

      Note. Use same quantity of each size.

      (2) *Torsion springs.*

Number	6.
Thickness	0.0950 in.
Length	16⅞ in.
Width	1 in.
Torque at 90 degree twist	275 ft.-lb.

(3) *Direct vision blocks.*

Number_____ 6.

Thickness_____ 2½ in.

b. TURRET TRAVERSING BALL RING.

(1) *Outer lower turret ring.*

Number of teeth_____ 364.

Width of tooth on pitch line_____ 0.262 in.

Pitch line from end of teeth_____ 0.133 in.

(2) *Turret traversing ball.*

Number of balls_____ 132.

Diameter of ball_____ 0.9995 to 1.0005 in.

c. TURRET TRAVERSING LOCK.   Plunger spring:

Free length_____ 0.960 in.

Pressure compressed to 1⅜ in_____ 19 to 21 lb.

d. PERISCOPE LINKAGE.

Outside diameter of bearings_____ 1.8499 to 1.8504 in.

Inside diameter of connector_____ 1.8502 to 1.8508 in.

Fit of bearing in connector_____ 0.0002 to 0.0009 in. (loose).

Inside diameter of bearings_____ 0.7870 to 0.7874 in.

Diameter of stud_____ 0.7875 to 0.7880 in.

Fit of bearing on stud_____ 0.0001 to 0.0010 in. (tight).

Width of snap ring groove in connector___ 0.068 to 0.078 in.

Thickness of snap ring_____ 0.064 to 0.066 in.

Clearance between ring and groove_____ 0.002 to 0.014 in. (loose).

Note. Wear Limit Column has been omitted in this chapter due to lack of necessity for same, however, clearances between parts and fits, as given, must be adhered to.

APPENDIX
REFERENCES

1. Publications Indexes

The following publications indexes should be consulted frequently for latest changes or revisions of reference given in this section and for new publications relating to matériel covered in this manual:

a. Ordnance supply catalog index_____ WD CAT ORD 2.

b. Ordnance major items and combina- SB 9-1.
tions, and pertinent publications.

c. List and index of War Department FM 21-6.
publications.

d. List of War Department films, film FM 21-7.
strips, and recognition film slides.

e. Military training aids_____ FM 21-8.

2. Standard Nomenclature Lists

a. VEHICLE.

Tank, light, M24; Carriage, motor, twin WD CAT ORD (*) SNL
40-mm. gun, M19. G-200.

Carriage, motor, 155-mm. howitzer, M41_ WD CAT ORD (*) SNL
G-236.

b. MAINTENANCE.

Antifriction bearings and related items__ WD CAT ORD 3 SNL
H-12.

Cleaning, preserving and lubricating ma- WD CAT ORD 3 SNL
terials; recoil fluids, special oils and K-1.
miscellaneous related items.

Lubricating equipment, accessories and WD CAT ORD 3 SNL
related dispensers. K-3.

Miscellaneous hardware_____ WD CAT ORD 3 SNL
H-2.

Oil seals_____ WD CAT ORD 3 SNL
H-13.

Soldering, brazing and welding materials, WD CAT ORD 3 SNL
gases and related items. K-2.

(*) See WD Catalog ORD 2 Index for published pamphlets of the Ordnance Supply Catalog.

Standard hardware	WD CAT ORD 5 SNL H-1.
Tools and supplies for ordnance base armament maintenance battalion.	WD CAT ORD 10 SNL N-315.
Tools and supplies for ordnance base automotive maintenance battalion.	WD CAT ORD 10 SNL N-325.
Tool-sets (common), specialists and organizational.	WD CAT ORD 6 SNL G-27 (sec. 2).
Tool sets (special), motor vehicles	WD CAT ORD 6 SNL G-27 (sec. 1).

c. ARMAMENT.

Gun, machine, caliber .30, Browning, M1919A4, fixed and flexible, M1919A5, fixed, M1919A6, ground mounts.	WD CAT ORD (*) SNL A-6.
Gun, machine, caliber .50, Browning, M2, Heavy Barrel, fixed and flexible; ground mounts.	WD CAT ORD (*) SNL A-39.
Gun, 75-mm., M5 and AN-M5A1; mount, gun, airplane, 75-mm., AN-M9.	WD CAT ORD (*) SNL C-60.
Howitzer, 155-mm., M1; Carriage, howitzer, 155-mm., M1A1, M1A2; mount, howitzer, 155-mm., M14 (T19).	WD CAT ORD (*) SNL C-39.

3. Explanatory Publications

a. FUNDAMENTAL PRINCIPLES.

Automotive electricity	TM 10-580.
Dictionary of United States Army terms.	TM 20-205.
Electrical Fundamentals	TM 1-455.
Military motor vehicles	AR 850-15.
Ordnance service in the field	FM 9-5.
Precautions in handling gasoline	AR 850-20.
Standard military motor vehicles	TM 9-2800.

b. OPERATION OF MATÉRIEL.

Light tank, T24 (M24)	TM 9-729.
155-mm. howitzer M1 and carriage M1 and M1A1.	TM-9-331.

c. MAINTENANCE AND REPAIR.

Basic Maintenance Manual	TM 38-650.
Cleaning, preserving, sealing, lubricating, and related matériels issued for ordnance material.	TM 9-850.

(*) See WD Catalog ORD 2 Index for published pamphlets of the Ordnance Supply Catalog.

Instruction guide care and maintenance TM 37–265.
of ball and roller bearings.

Maintenance and care of pneumatic tires TM 31–200.
and rubber treads.

Ordnance maintenance: Carburetor TM 9–1826A.
(Carter).

Ordnance maintenance: Electrical equip- TM 9–1825A.
ment (Delco-Remy).

Ordnance maintenance: Hydraulic trav- TM 9–1731G.
ersing mechanism (Oil Gear) for medi-
um tanks M4 and modifications.

Ordnance maintenance: Light tanks M5, TM 9–1729A.
M5A1, and M24, 75-mm. howitzer
motor carriage M8, and twin 40-mm.
gun motor carriage M19: Engines,
cooling systems and fuel systems.

Ordnance maintenance: Light tank M24 TM 9–1729B.
and twin 40-mm. gun motor carriage
M19 Transmission, transfer unit, propeller shafts, controlled differential
and final drives.

Ordnance maintenance: Motor vehicle in- TM 37–2810.
spection and preventive maintenance
services.

Ordnance maintenance: Speedometers, TM 9–1829A.
tachometers, and recorders.

Ordnance maintenance: 155-mm. howit- TM 9–1331.
zer M1, 4.5-inch gun M1 and carriages
M1 and M1A1.

Ordnance maintenance: Vehicular main- TM 9–1834A.
tenance equipment: Grinding, boring,
valve reseating machines and lathes.

d. PROTECTION OF MATÉRIEL.

Camouflage, basic principles_____ FM 5–20.
Chemical Decontamination Company____ FM 3–70.
Decontamination___ _____ TM 3–220.
Decontamination of armored force ve- FM 17–59.
hicles.
Defense against chemical attack_____ FM 21–40.
Explosives and demolitions_____ FM 5–25.
Military chemistry and chemical agents__ TM 3–215.

e. STORAGE AND SHIPMENT.

Ordnance company, depot_____ FM 9–25.
Ordnance packaging and shipping_____ TM 9–2854.

Ordnance storage and shipment chart, SB 9–OSSC–G.
group G—Major items.

Preparation of unboxed ordnance maté- SB 9–4.
riel for shipment.

Registration of motor vehicles_____ AR 850–10.

Rules governing the loading of mecha-
nized and motorized Army equipment,
also major caliber guns, for the United
States Army and Navy, and open top
equipment published by Operations and
Maintenance Department of Associa-
tion of American Railroads.

Storage of motor vehicle equipment_____ AR 850–15.

INDEX

U. S. GOVERNMENT PRINTING OFFICE O—1947

www.ingramcontent.com/pod-product-compliance
Lightning Source LLC
Chambersburg PA
CBHW060614200326
41521CB00007B/773